atical Research · Mathematische Forschung

haftliche Beiträge
geben von der
der Wissenschaften der DDR
r Mathematik

0

ets Applications,
ological Approaches,
sults

Methodologi

Mathem

Wissensc
herausge
Akademi
Institut fü

Band 3

Fuzzy S
Methoc
and Re

Fuzzy Sets Applications, Methodological Approaches, and Results

Proceedings of the International Workshop on Fuzzy Sets Applications cosponsored by IIASA and convened in cooperation with IAEA held on the Wartburg, Eisenach (GDR) at March 3–8, 1985

---

edited by

Steffen Bocklisch, Sergei Orlovski,
Manfred Peschel, and Yasushi Nishiwaki

Akademie-Verlag Berlin 1986

Herausgeber:

Dr. Steffen Bocklisch
Akademie der Wissenschaften der DDR
Institut für Mechanik

Dr. Sergei Orlovski
International Institute for Applied
Systems Analysis (IIASA)

Prof. Dr. Manfred Peschel
Akademie der Wissenschaften der DDR
Forschungsbereich Mathematik/Informatik

Prof. Dr. Yasushi Nishiwaki
International Atomic Energy Agency (IAEA)
Risk Assessment Programme

Die Titel dieser Schriftenreihe werden vom Originalmanuskript der Autoren reproduziert.

ISSN 0138 - 3019
ISBN 3-05-500016-1

Erschienen im Akademie-Verlag Berlin, DDR-1086 Berlin, Leipziger Straße 3-4

Lizenznummer: 202 · 100/555/85
Printed in the German Democratic Republic
Gesamtherstellung: VEB Kongreß- und Werbedruck, 9273 Oberlungwitz
Lektor: Dr. Reinhard Höppner
LSV 1095
Bestellnummer: 763 477 2 (2182/30)

03000

Preface

This volume contains the papers of the International Workshop on Fuzzy Sets Applications held on the Wartburg, Eisenach (GDR), March 3-8, 1985. The workshop was organized by the Central Institute of Cybernetics and Information Processes of the Academy of Sciences of the GDR and was co-sponsored by the International Institute for Applied Systems Analysis (IIASA) and convened in cooperation with the International Atomic Energy Agency (IAEA). Some additional tutorial papers are included in order to give the newcomer an introduction in fuzzy thinking.

The papers included in these proceedings were not formally reviewed and the authors are completely responsible.

The workshop represents the first scientific event with international participation and support of the GDR working group "Fuzzy Sets and Systems" (directed by Dr. St. Bocklisch, Institute for Mechanics of the Academy of Sciences of the GDR).

The intention of the workshop was to make the reader familiar with recent methodological developments and at the same time they demonstrate the productivity of GDR scientists in this field. Based on the long-standing tradition of the seminar "Polyoptimization and Fuzzy Model-building", practized since 1972 under the leadership of Prof. Dr. M. Peschel at the Academy of Sciences of the GDR, some sound application routes have been formed which were presented at the workshop.

The fundamental methodological problems are naturally related to the thinking itself. As it is well-known, fuzzy thinking is closely connected with human thinking and action and is breaking with the two-valued black-white logic. At the same time fuzzy logic represents a starting point for utilization of the fuzzy way of thinking in the field of artificial intelligence, especially for expert systems. First of all, artificial intelligence is human intelligence transformed into technical capacity.

Based on fuzzy logic and fuzzy set theory, the fuzzy systems theory is obliged to mend the other basic mathematical branches also with their own specific instruments. This process is in the very beginning, yet. First steps have been made towards algebra and analysis and there are attempts to separate fuzzy systems theory as a possibility theory from the probability theory. A number of papers of this workshop is devoted to this aim. The essential application directions of the fuzzy method naturally aim at model-building and decision-making as the two fundamental columns of every systems theory.

Fuzzy decision-making is intrinsically multicriterial. Due to this fact, a fuzzy notion of quality is normally decomposed and divided into a finite number of 'less fuzzy' quality attributes (quality criteria) for better handling.

As a fuzzy alternative to usual Operations Research techniques the fuzzy approach contributes much with special methods to decision-making in general, but also to control strategies for industrial applications.

Even at the international level fuzzy model-building is very closely connected with qualitative model-building in virtue of classification, although there are contributions to the determination of qualitative input-output models, too.
Therefore, the main emphasis of this workshop in model-building was stressed on classification and structure-building and less on general questions of model-building.
Within a round-table-discussion during the workshop an interesting contribution to the possibilities of fuzzy evaluations in risk situations of nuclear energetics was represented.
These proceedings give the reader a very good insight into foundations and recent developments in the field of fuzzy methodology with a strong orientation to applications, above all towards supporting means of diagnosis.
The software contained in the program packages (FORTRAN 77) is accessible to a broad utilization, a number of codes is available in BASIC for microcomputer applications, too.

We would like to thank our partners of IIASA and IAEA for good cooperation, and the Akademie-Verlag Berlin, especially Dr. Höppner and Ms. Reiher, for the quick and flexible assistance in preparation of these proceedings.

July 1985 Manfred Peschel

CONTENTS

BASIC METHODOLOGICAL PROBLEMS

Kóczy, L. T.
Interactive formulation of fuzzy algebra and its application as a model of function capability 9

Novák, V.
The origin and claims of fuzzy logic 21

Novák, V.
First-order fuzzy logic and its applications in fuzzy modelling 27

Wagenknecht, M., and K. Hartmann
On the solution of direct and inverse problems for fuzzy equation systems with tolerances 37

Gottwald, S.
On some theoretical problems concerning the construction of fuzzy controllers 45

Fedrizzi, M., and J. Kacprzyk
Essentials of fuzzy sets theory: a linguistic perspective 56

Kacprzyk, J.
Fuzzy dispositions as a tool for more human-perception-consistent models in systems analysis 64

FUZZY DECISION - MAKING

Böttner, R.
Fuzzy approach to multiple criteria decision-making 74

Werners, B.
A flexible mathematical programming system 86

Ramík, J.
An application of fuzzy optimization to optimum allocation of production 96

Khurgin, J.I., and V.V. Polyakov
Stability in the problem of multicriterial fuzzy decision-making 105

Khurgin, J.I., and V.V. Polyakov
Fuzzy analysis of the group concordance of expert preferences, defined by Saaty matrices 111

Khurgin, J.I., and V.V. Polyakov
Fuzzy approach to the analysis of expert data 116

FUZZY MODEL-BUILDING

Orlovski, S.A.
Fuzzy goals and sets of choices in two-person games 125

Straube, B.
Model building and fuzzy systems 133

Bocklisch, S.F., and U. Priber
A parametric fuzzy classification concept 147

Peschel, M., B. Straube, and W. Mende
Fuzzy inferences for the analysis of qualitative behaviour 157

FUZZY STRUCTURE FORMATION

Straube, B.
Some remarks to fuzzy cluster analysis 165

Straube, B.
Fuzzy cluster analysis by means of FDA 175

Voigt, H.-M.
Fuzzy structure formation based on natural selection 181

FUZZY APPLICATIONS

Lipp, H.-P., and R. Guenther
An application of a fuzzy Petri net in complex industrial systems 188

Weise, A., and W. Kühnel
Application of a fuzzy set concept to the design of chemical syntheses 197

Bocklisch, S.F., F. Bilz, and W. Schüler
A fuzzy classification diagnosis system 203

Nishiwaki, Y.
Possible application of fuzzy set theory to nuclear safety analysis, risk perception and nuclear plant siting evaluation 213

Lederman, L.
Summary of the panel discussion on the use of fuzzy set theory to nuclear safety and risk assessment 232

INTERACTIVE FORMULATION OF FUZZY ALGEBRA AND ITS APPLICATION AS A MODEL OF FUNCTION CAPABILITY

László T. Kóczy [1/]

Introduction

Interactive formulation of fuzzy algebra has a history almost as old as traditional distributive lattice /max-min/ formulation. After some predecessors in the field of /non-fuzzy/ multivalued logics /as Reichenbach and Bochvar/, in his very first papers Zadeh, creator of the idea of fuzzy sets already suggested an alternative way for defining fuzzy operations /"soft" operations or product and sum, [10]/, although he was not really concerned about the properties of the structure of this construction. As these operations did not posess distributive properties, most of the authors neglected them totally, although they later proved to be useful in many applications. We started our investigations concerning the interactive operations in 1973 and published some first results in 1974 resp. 1975 [4], [5]. Some further papers on more precise and far more general ideas must also be mentioned [8], [1], [6] and [7]. Also in 1975 some parallel results were published by Rödder and Hamacher [9], [2]. These latter results will be discussed in a somewhat modified form also in this paper. In recent years some more general results were achieved showing a much general class of operations /t- and s-norms/, especially by Klement [3], we will not discuss however such broad aspects of interactive structure formulation.

Interactive fuzzy algebra

There are many ways of construction for an axiomatic system in accordance with the original sum and product operations, we refer to the above mentioned publications of the author. In this paper we shall use as follows:

Let

$$R = \langle P, +, *, d, \geq \rangle$$

be a fuzzy range, where P is a set ordered by $\geq$, so that P contains its maximum /1/ and minimum /0/, i.e. $0, 1 \in P$ and $0 \leq p$ and $1 \geq p$ for all

1/
Department of Communication Electronics,
Technical University of Budapest,
Budapest, Stoczek u. 2., 1111, Hungary

p ∈ P. +, * and d are binary operations over which P is closed, with following properties /in the next = has the meaning ≤ and ≥ , while < stands for ≤ but not ≥ etc./:

1/ +/a,b/ = +/b,a/
*/a,b/ = */b,a/
commutativity

2/ +/+/a,b/,c/ = +/a,+/b,c//
//a,b/,c/ = */a,*/b,c//
associativity

3/ +/a,0/ = a
*/a,1/ = a
universal bouds

4/ +/*/a,b/,*/a,c// ≥ */a,+/b,c//
and > iff
+/*/a,b/,*/a,c// > 0 and
*/a,+/b,c// < 1.
/+/a,b/,+/a,c// ≤ +/a,/b,c//
and < iff
*/+/a,b/,+/a,c// < 1 and
+/a,*/b,c// > 0
distributive <u>inequalities</u>

5/ +/a,b/ ≥ a
and > iff
a < 1 and b > 0
*/a,b/ ≤ a
and < iff
a > 0 and b < 1
monotonicity

6/ +/a,1/ = 1
*/a,0/ = 0
zero properties

7/ d/a,a/ = 0
self distance

8/ d/a,b/ ≤ d/a,c/ iff a ≤ b ≤ c
and < iff b < c
distance monotonicity

9/ d/1,0/ = 1
norming

10/ d/1,d/1,a// = a
involution

11/ d/1,+/a,b// = */d/1,a/,d/1,b//
d/1,*/a,b// = +/d/1,a/,d/1,b//
De Morgan's laws

12/ d/a,b/ = d/d/1,a/,d/1,b//

distance symmetry

13/ +/a,x/ = b has exactly one solution for x if a < b

ж/a,x/ = b has exactly one solution for x if a > b

It is clear that this system is not without redundancy, i.e. many of the properties are depending on others. Let us take an example:

The first part of 11/ is

d/1,+/a,b/ = ж/d/1,a/,d/1,b//

Let a = d/1,x/ and b= d/1,y/.

Then we have

d/1,+/d/1,x/,d/1,y/// = ж/d/1,d/1,x/,d/1,d/1,y///

Using 10/ the above is equivalent with

d/1,+/d/1,x/,d/1,y/// = ж/x,y/

Now we are going to calculate

d/1,ж/x,y// using /1/.

d/1,ж/x,y// = d/1,d/1,+/d/1,x/,d/1,y//// = +/d/1,x/,d/1,y//

/Because of 10/./

So we obtained the second part of 11/.

The most interesting among the axiomes are those under 13/ which provide a way for inverting "+" and "ж".

Based on R, the fuzzy structure is defined by:

$$S = \langle F,R,M,V,\wedge,^{-} \rangle$$

where F is a set /universe/ and M is an isomorphism between F and P:

$$M: F \leftrightarrow P,$$

and the operations V, $\wedge$ and $^{-}$ are defined over F as follows:

$$A \vee B = M^{-1}/+/M/A//,M/B///$$

$$A \wedge B = M^{-1}/ж/M/A//,M/B///$$

$$\bar{A} = M^{-1}/d/M/1/,M/A///$$

if $A,B \in F$.

An only technical generalization of the above is taking $\Pi = \prod_{i=1}^{n} P_i$, where for P_i the axiomes are valid and

$$R^n = \langle \Pi, +^n, ж^n, d^n \rangle,$$

so that e.g.

$$+/a^n,b^n/ = \begin{bmatrix} +_1/a_1,b_1/ \\ \dots \\ +_n/a_n,b_n/ \end{bmatrix}$$

if $a^n,b^n \in \Pi$ and $a^n = \begin{bmatrix} a_1 \\ \vdots \\ a_n \end{bmatrix}$ etc.

There are some very important properties of the inverse operations V^i

and Λ^i:

A/ $A \vee /B\vee^i C/ = /A\vee B/\vee^i C$

$A \wedge /B\wedge^i C/ = /A\wedge B/\wedge^i C$

Proof:

We prove

$+/a,x/ = y$ /2/

where $+/c,x/=b$, $+/c,y/=+/a,b/$.

Calculate $+/c,+/a,x//$, which is $+/a,+/c,x//$ because of 1/ and 2/ i.e. $+/a,b/$. But then using 13/ we have /2/. Q.e.d.

B/ $/A \vee^i C/\vee B = A\vee/B\vee^i C/$

$/A \wedge^i C/\wedge B = A\wedge/B\wedge^i C/$

Proof:

Because of 1/, B/ can be written as $B\vee/A\vee^i C/ = A\vee/B\vee^i C/$, etc. Using A/, both sides are $/A\vee B/\vee^i C$, etc. Q.e.d.

C/ $A\vee^i/B\vee^i C/ = /A\vee^i B/\vee C$

$A\wedge^i/B\wedge^i C/ = /A\wedge^i B/\wedge C$

Proof:

We prove

$y = +/z,c/$ /3/

where $+/x,y/=a$, $+/c,x/=b$ and $+/b,z/=a$.

$+/x,+/z,c//=+/+/x,c/,z/=+/b,z/=a$, but as $+/x,y/=a$ and y is unique, we obtained /3/. Q.e.d.

D/ $A \vee^i /B\vee C/ = /A\vee^i B/\vee^i C$

$A \wedge^i /B\wedge C/ = /A\wedge^i B/\wedge^i C$

Proof:

We prove

$x = z$

where $+/+/b,c/,x/=a$ and $+/c,z/=y$, so that $+/b,y/=a$.

Let us calculate

$+/+/b,c/,z/=+/b,+/c,z//=+/b,y/=a$, but then $x=z$, because of 13/. Q.e.d.

As a matter of course $A\vee^i B \neq B\vee^i A$ as if $A\vee^i B$ exists, nothing can be said about the existance of $B\vee^i A$, etc.

By properties A/ to D/ calculations using the inverse operations can be done very conveniently.

Representations

It is interesting to see now, what concrete operations fulfil our axiomes, which also can be used for practical purposes. Usually $P=[0,1]$, then product and "sum" are suitable for $*$ and $+$, while d can be represented by subtraction:

$$M/\bar{A}/ = 1-M/A/$$
$$M/A\vee B/ = M/A/+M/B/-M/A/\cdot M/B/ \qquad /4/$$
$$M/A\wedge B/ = M/A/\cdot M/B/.$$

We are going on using m_A instead of $M/A/$, etc.

A much more general class for $*$ and $+$ is given by

$$m_{A\vee B} = \frac{/m_A+m_B-m_A\cdot m_B/\cdot c_1+m_A\cdot m_B/c_1-c_2/}{m_A\cdot m_B/c_1-c_2/+c_1}$$
$$m_{A\wedge B} = \frac{m_A\cdot m_B\cdot c_1}{/m_A+m_B-m_A\cdot m_B//c_1-c_2/ + c_2}$$

where $c_1 > 0$, $c_2 > 0$, $c_1 \geq c_2-1$ or $c_2 < 0$, $c_1 \leq c_2-1$ [9].
Negation $/\bar{A}/$ is the same as in /4/.

The inverse operations are in the first case:

$$m_{A\vee i_B} = \frac{m_A-m_B}{1-m_B}$$
$$m_{A\wedge i_B} = \frac{m_A}{m_B}$$

and in the general case:

$$m_{A\vee i_B} = \frac{/m_A-m_B/c_1}{/1-m_B/\cdot c_1-m_B/1-m_A/\cdot c_2}$$
$$m_{A\wedge i_B} = \frac{m_A\cdot//1-m_B/c_2+m_B\cdot c_1//}{m_B\cdot c_1-m_A/1-m_B//c_1-c_2/}$$

which expressions always exists if $m_B < m_A/\leq 1/$ and $m_B > m_A/\geq 0/$, resp.

Two very simple examples for /4/

1/ If $m_A=0.2$, $m_B=0.5$:

$m_{\bar{A}}=0.2$

$m_{A\vee B}=0.6$ $\qquad$ $m_{B\vee i_A} = 0.375$

$m_{A\wedge B}=0.1$ $\qquad$ $m_{A\wedge i_B}=0.4$

2/ If $\underline{m}_A=\begin{bmatrix}0.1\\0.3\end{bmatrix}$, $\underline{m}_B=\begin{bmatrix}0.2\\0.8\end{bmatrix}$ $\quad$ /n=2/:

$$\underline{m}_{\bar{A}}=\begin{bmatrix}0.9\\0.7\end{bmatrix}$$

$$m_{A\vee B}=\begin{bmatrix}0.28\\0.86\end{bmatrix}$$

$$m_{A\wedge B}=\begin{bmatrix}0.02\\0.24\end{bmatrix}$$

Large telecommunication equipment

In the telecommunications, especially in switching, there are very large real time equipment with a modular structure where it seems to be reasonable to introduce fuzzy degrees for function capability.

These equipment have a few very important properties:
- large dimensions, which induce a large amount of component modules
- high reliability requirements /as e.g. 1h/20 years allowed maximal total stillstand time/, meaning a redundant structure
- possibility of steady changing, also under operation, which means as well the possibility for quantitative development as for qualitative alterations

These three features cause the appearance of quite complicated hierarchical structures which is especially interesting in the multiprocessor control system.

As maybe the most typical example for such equipment we mention a public network telephone exchange. Dimensions can go until 100 or 120 thousand subscribers and more 10 thousands of other terminals /trunks, tie-lines, etc./. These exchanges must work 24 hours a day with the above mentioned very strict reliability criterion. During the 20 or 30 years of operation a permanent growth in the number of subscribers, trunks, etc., in general the switched traffic and the constant need for a quick evolution of services provided by the exchange result in a steady change of the anyway complicated structure.

Another important feature is that partly functioning units must also be considered, as e.g. if one of ten elements having identical functions stops working, the remaining set works on but with less capability. Considering a theoretical maximal traffic to be handled at a certain point, the real system built in will be able to deal with only a part of it. It is never necessary to calculate with the maximum as the probability of its occuring is usually very low /except some cases of constant traffic/. So the function capability of any real component will be well under 1. Because of reliability and other reasons, instead of having one such element /e.g. a microprocessor/ with, let us say, 0.6 function capability degree, more elements each with a lesser degree are used. So different systems with different traffic handling capacity can be built of the same component /using less or more pieces of it/.

Next we shall summarize some properties of such systems:
1/ Adding two components with identical functions /in load sharing/ the resulting system will work independently from the, sequence of the two

subsystems' putting into operation.

If two subsystems with different tasks are applied, usually the resulting system works in a degree independent from the inner sequence of performing the two functions.

2/ If a homogenous interconnection is used putting more than two subsystems together, the resulting system will be independent in the function capability from grouping or sequence of the subsystems.

3/ Adding a non-functioning unit the system's function capability will not be changed. Adding a unit with a new but necessary function, which unit works with absolutely sufficient capacity, etc. The function capability of the resulting system is inaffected.

4/ Parallelly doubling any functional unit increases, serially doubling any unit decreases the function capability, assuming the original system was not totally failing to work or with absolutely sufficient capacity.

5/ Adding a new unit in load sharing increases the degree if the original was of less than absolutely sufficient and the new one of more than totally zero capacity. Adding a new but necessary function, with the according unit /having less than completely sufficient capacity/ the result degree will decrease, if the original worked at all.

6/ Adding a completely sufficient unit in load sharing, the result will also be completely sufficient, independently from the original subsystem. Adding a new function however executed by a totally wrong unit, the system will not work under the new conditions.

7/ If a part of a load-sharing set is taken out, the original and resulting degree will define uniquely the function capability of the element taken out, assuming the degree decreases at all. If a function is omitted - with according functional unit too - the original and resulting degrees will define uniquely the function capability, assuming the degree increases at all.

The above quite general statements are fully in accordance with axiomes 1/ to 6/ and 13/ if load sharing is considered as V and function sequence /series/ is considered as $\wedge$, 0 meaning totally wrong and 1 the absolutely sufficinet elements function capability.

If there are more then one function features /e.g. traffic handling capacity,maintainability/,the degrees can be given in vectors, so we obtain the R^n case.

Characteristic structural elements

In this part we show a series of examples for typical substructures occuring in such systems.

1/ Single module

It is described by a vector of degrees

$$\underline{m} = \begin{bmatrix} m_1 \\ \vdots \\ m_n \end{bmatrix},$$

usually $0 < m_i < 1$. $m_i = 0$ is used in the case of failure or switch-off, $m_i = 1$ in the case of very simple elements /e.g. a single switch/.

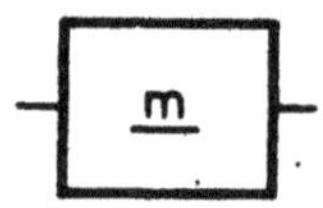

2/ Serial functions /irredundant construction/

It is described by vectors $\underline{m}_1, \dots \underline{m}_k$; $\underline{m}_i = \begin{bmatrix} m_{i1} \\ \vdots \\ m_{in} \end{bmatrix}$

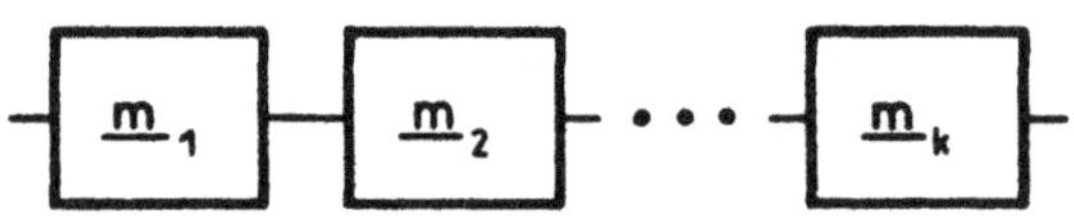

As illustration we shall use in the next examples the product and su representation for calculating resulting degrees - although we state only that "greater" or "less" relations will be correct, not the actual degrees in any exact meaning. So the system degree $\underline{m}_s$ will be:

$$\underline{m}_s = \pi/\underline{m}_1, \dots, \underline{m}_k/ = \prod_{i=1}^{k} \underline{m}_i = \begin{bmatrix} \prod_{i=1}^{k} m_{i1} \\ \vdots \\ \prod_{i=1}^{k} m_{in} \end{bmatrix}$$

If for any j $0 < m_{ij} < 1$ for all i, $0 < m_{sj} < 1$. If $m_{ij} = 0$ for at least one i, $m_{sj} = 0$. If to a given system a new necessary function is added,

$m'_{sj} < m_{sj}$ if $m_{k+1,j} < 1$. $m_{k+1,j}=1$ does not affect the degree in property j.

3/ <u>Load sharing of more elements</u>

The elements are described by vectors $\underline{m}_1,\dots,\underline{m}_k$.

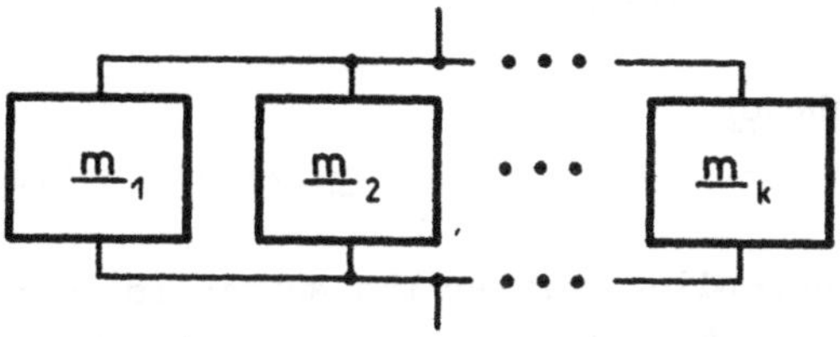

$$\underline{m}_s = +/\underline{m}_1,\dots,\underline{m}_k/ = \sum_{i=1}^{k} \underline{m}_i - \sum_{i=1}^{k} \sum_{j\neq i} \underline{m}_i\underline{m}_j + -\dots/-1/^{k-1} \prod_{i=1}^{k} \underline{m}_i$$

In the usual case $\underline{m}_i$ is constant $\underline{m}$. Then

$$\underline{m}_s = \sum_{i=1}^{k} /-1/^{i-1}\cdot/{k \atop i}/\cdot\underline{m}^i$$

This can be proven easily, as for k=1 it holds:

$\underline{m}_s = /-1/^{1-1}\cdot/{1 \atop 1}/\cdot\underline{m}^1 = \underline{m}^1$,

and assuming now that it holds for k, taking a /k+1/th unit,

$$\underline{m}'_s = \underline{m}_s + \underline{m} - \underline{m}_s \cdot \underline{m} =$$

$$= \sum_{i=1}^{k} /-1/^{i-1}\cdot/{k \atop i}/\underline{m}^i + \underline{m} - \underline{m}/\sum_{i=1}^{k} /-1/^{i-1}\cdot/{k \atop i}/\cdot\underline{m}^i/ =$$

$$= \sum_{i=1}^{k} /-1/^{i-1}\cdot/{k \atop i}/\underline{m}^i + /-1/^{o}\cdot/{k \atop o}/\cdot\underline{m}^1 + \sum_{j=2}^{k+1} /-1/^{j-1}/{k \atop j-1}/\cdot\underline{m}^j =$$

$$= /-1/^{o}//{k \atop o}/+/{k \atop 1}//\cdot\underline{m}^1 + \sum_{i=2}^{n} /-1/^{i-1}\cdot//{k \atop i-1}/+/{k \atop i}//\cdot\underline{m}^i + /-1/^{k+1-1}\underbrace{/{k \atop k}/}_{=/{k+1 \atop k+1}/}\cdot\underline{m}^{k+1} =$$

$$= \sum_{i=1}^{k+1} /-1/^{i-1}\cdot/{k+1 \atop i}/\cdot\underline{m}^i. \text{ Q.e.d.}$$

It is obvious, that $0 < \underline{m}_s < \underline{m}'_s < 1$ if $0 < \underline{m}_i < 1$ for all i. $\underline{m}_i=1$ means that all other units are superfluous, so load sharing is not used. What happens if the ith unit gets a defect? The reduced system's function capability will be:

$$\underline{m}''_s = \underline{m}_s V^{i}\underline{m}_i = \frac{\underline{m}_s - \underline{m}_i}{1-\underline{m}_i} < \underline{m}_s$$

/$\underline{m}_i$ must have been $>$ 0 before the defect/.

In the homogenous case

$$\sum_{i=1}^{k} /-1/^{i-1}./{k \atop i}/.\underline{m}^i \rightarrow \sum_{i=1}^{k-1} /-1/^{i-1}./{k-1 \atop i}/.\underline{m}^i.$$

We omit the not very difficult proof.

4/ Standby with switchover

It is described by the switches $\underline{s}_1/t/,\ldots,\underline{s}_k/t/$ and units $\underline{m}_1,\ldots,\underline{m}_k$.

Usually $\underline{s}_i/t/=1$ or 0 depending whether the switch is on or off. At given time t exactly one of the k switches assumes 1 all other are 0-s.

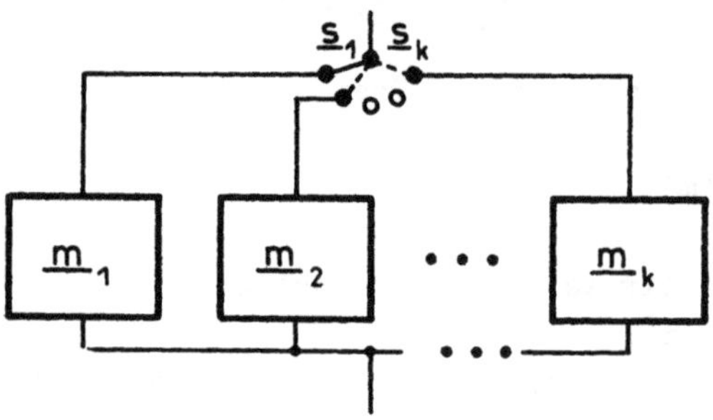

The formula can be obtained by using 2/ and 3/ together. One parallel "way" is described by $\times/\underline{s}_i,\underline{m}_i/=\underline{s}_i\cdot\underline{m}_i$. They can be considered as working in load sharing, so

$\underline{m}_s = +/\underline{s}_1\cdot\underline{m}_1,\underline{s}_2\cdot\underline{m}_2,\ldots,\underline{s}_k\cdot\underline{m}_k/ = \underline{s}_i\cdot\underline{m}_i = \underline{m}_i$ where $\underline{s}_i=1$.

If the working unit gets a defect, $\underline{m}_i$ changes to 0, i.e. $\underline{m}_s$ also. After switchover a new switch $\underline{s}_j$, $j\neq i$ assumes 1, so $\underline{m}_s$ will be $\underline{m}_j$. If we define $xV^1x=0$ $/x\neq 1/$, we can use $\frac{x-x}{1-x}=0$ for calculating the result:

$\underline{m}_s/\text{defect}/ = \underline{m}_s V^1 \underline{m}_i = 0.$

If any other element fails to work, it cannot be detected through $\underline{m}_s$ as the according $\underline{s}_j$ is 0. If the active switch gets wrong $\underline{m}_s$ also changes to 0, and the same recovery fenomenon appears.

In the case of standby structures we must stress, that these considerations are valid only from the point of view of traffic handling capability. If the reliability of the system is calculated, the same structure transforms into case 3/ as the standby elements guarantee a higher probability of the complete system functioning. Then $\underline{m}_{ij}$ can be e.g. the probability of correct working of the ith unit /p/, i.e. assuming one standby with the same probability we have $m_s = 2m - m^2 = 2p - p^2$ for the reliability of the system. This is in accordance with the probability of correct functioning: $1 - /1-p/^2$, assuming independence between the two units. This example also shows that certain features induce another model structure, so the vectorial approach is not at all usable in many cases.

Finally we show an example for a /simplified/ real exchange structure:

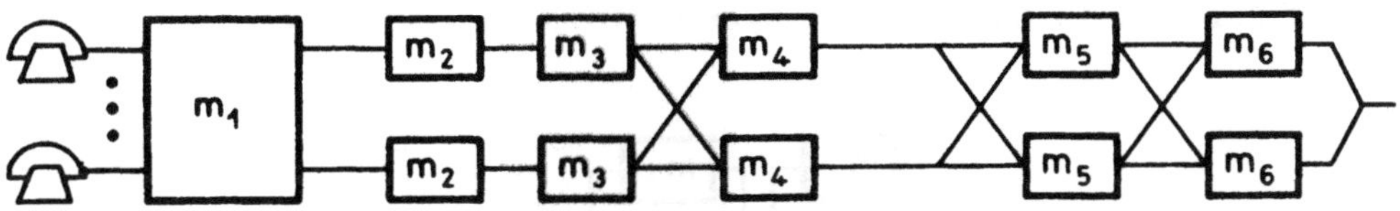

Here the m_2-m_3 pairs work in standby, m_4 in load-sharing, m_5 and m_6 also in standby but separately. The model structure is:

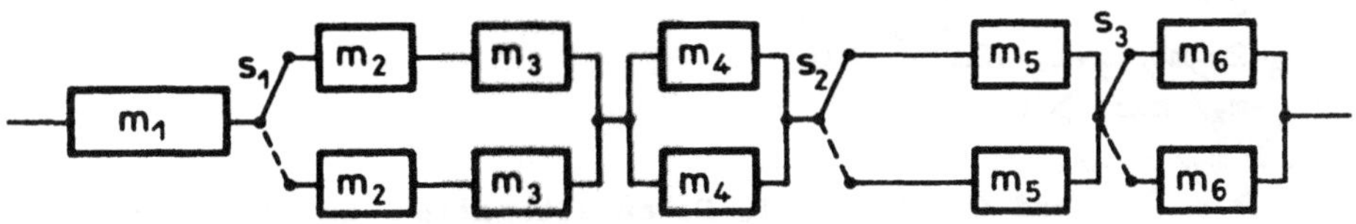

The resulting function capability is:

$$m_s = m_1 \cdot /\underset{\substack{\uparrow \\ s_1}}{1} \cdot m_2 \cdot m_3/ \cdot /2m_4 - m_4^2/ \cdot /\underset{\substack{\uparrow \\ s_2}}{1} \cdot m_5/ \cdot /\underset{\substack{\uparrow \\ s_3}}{1} \cdot m_6/$$

The reliability model structure is, with both m_4 working:

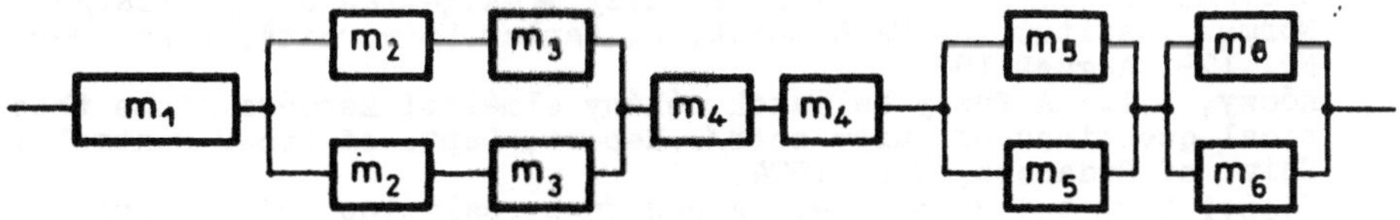

Reliability degree is:

$$m_s' = m_1 \cdot /2m_2 \cdot m_3 - m_2^2 \cdot m_3^2/ \cdot m_4^2 \cdot /2m_5 - m_5^2/ \cdot /2m_6 - m_6^2/$$

Using the model some conclusions can be obtained. The structure used in the case of m_2-s and m_3-s is more convenient than a structure like this:

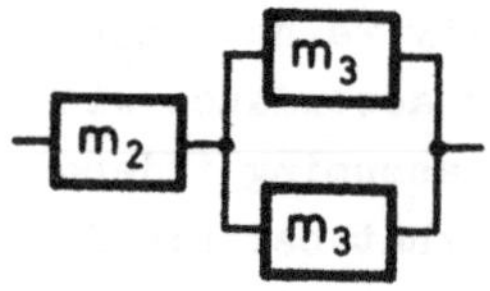

as $m_2 \cdot /2m_3-m_3^2/ < 2m_2 \cdot m_3-m_2^2m_3^2$

$$m_2^2m_3^2 < m_2m_3^2$$

if $0 < m_2$, $m_3 < 1$. However, yet better is such a configuration:

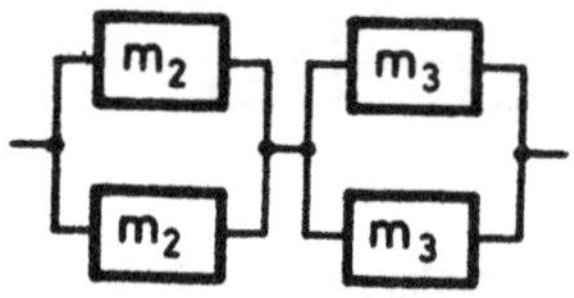

as

$/2m_2-m_2^2//2m_3-m_3^2/ > 2m_2m_3-m_2^2m_3^2$

$2m_2m_3-2m_2m_3/m_2+m_3/+2m_2^2m_3^2 > 0$

$1-/m_2+m_3/+m_2m_3 > 0$

$1-m_2 > m_3/1-m_2/$,

but $1 > m_3 > 0$ and $0 < 1-m_2 < 1$. So an alteration in the structure would bring an increase of reliability.

<u>References</u>

[1] Győrfi, L.and Kóczy, L.T.: An Algorithm for Nonparametric Decision Motivated by Fuzzy Approach. Problems of Control and Information Theory, <u>8</u> /1979/, pp.229-238.

[2] Hamacher, H.: Über logische Verknüpfungen unscharfer Aussagen und deren zugehörige Bewertungsfunktionen, Arbeitsbericht. Lehrstuhl für Unternehmensforschung RWTH Aachen, Juli 1975.

[3] Klement, E.P.: Construction of Fuzzy σ-Algebras Using Triangular Norms. Institut für Mathematik, J. Kepler Universität Linz, Bericht Nr. 179, August 198o.

[4] Kóczy, L.T.: A fuzzy halmazok néhány elméleti kérdése /Some theoretical questions of fuzzy sets/. Report, Dept. of Proc. Control,Tech. Univ. of Budapest, Jan. 1974.

[5] Kóczy, L.T. and Hajnal, M.: A new fuzzy calculus and its application as a pattern recognition technique, in: Modern Trends of Cybernetics and General Systems, Proc. of the 3rd Congress of WOGSC, Bucharest, Aug. 1975, Springer, Berlin etc., pp. 66-81.

[6] Kóczy, L.T.: Vectorial I-fuzzy Sets, in: Approximate Reasoning in Decision Analysis, North-Holland, Amsterdam etc., pp. 151-156.

[7] Kóczy, L.T. and Hajnal, M.: Classification of Textures by Vectorial Fuzzy Sets, <u>ibidem</u>, pp. 157-164.

[8] Kóczy, L.T.: Interactive σ-algebras and Fuzzy Objects of Type <u>N</u>, Journal of Cybernetics, <u>8</u> /1978/, pp. 273-290.

[9] Rödder, W.: On "and" and "or" connectives in fuzzy set theory, Working paper, Lehrstuhl für Unternehmensforschung RWTH Aachen,Jan.1975

[10] Zadeh L.A.: Fuzzy Sets, Information and Control, <u>8</u> /1965/,pp.338-353

THE ORIGIN AND CLAIMS OF FUZZY LOGIC

Vilém Novák +

What is fuzzy logic ?

The snow is white - similar ordinary utterances we use very often in our everyday reasoning. Is it true or not? One may immediately answer "yes". But what does it mean "white"? Is the snow white in an industrial area or near of it? In this case we might rather say: "it is almost white", it is white but not very much", etc. The notion "white" is vague, or, if you want, fuzzy. If one asks us about the truth of the proposition, we may answer e.g. "it is true in 80%", it is half true", etc. We see that it leads us to using degrees of truth. Our human reasoning uses the logic having more truth values than "true" and "false".

Many-valued logic is studied since thirties. Fuzzy sets theory helped to enliven the interest in it, brought new inspiration,concepts and motivation. Therefore , it is now usual to call a logic having more than two truth values the fuzzy logic. In fact, the term "fuzzy logic" has two basic meanings:

- many valued logic
- linguistic logic

The second one uses words such as "true, very true, rather false" etc. as truth values. It will be discused later.

Human thinking can be modelled using fuzzy logic

When thinking, man uses natural language. Specific feature of the natural language is vagueness of its meaning. It follows that human thinking uses vague notions as its organic component. It does not mean that human thinking is incorrect. On the contary, the possibility to work with vague notions gives human thinking its power and shares probably also in its creativity. On the other side, non-ability to work with vague notions on the computers is one of the reasons that their work is still very far from human thinking. It resembles more to mechanics repetition of simple acts than to real human thinking.If we want to make computers to think like people we must search a suitable mathematical tool by means of which such thinking could be modelled.This tool can be fuzzy logic. Fuzzy logic, especially its many-valued part, is now a powerful theory with deep structure which is flexible enough

----

+ Czechoslovak Academy of Sciences, Mining Institute, Hladnovská 7
710 00 Ostrava , Czechoslovakia

to cover many properties of the natural language in connection with human thinking. The copmplexity of the natural language is tremendous. Therefore we cannot be sure that our tool (regardless what kind it is) will meet all the finenesses in our mind. We can, however, be more or less sure that it grasps the essence of its which being implemented on the computer will make his performance resemble to human´s (excluding any kind of emotion, of course).

The basic laws of rigorous thinking are formulated in the classical two-valued logic. It is very well developed and deep theory according to which our contemporary computers behave .

If one wants to make irrefutable inference be must use cassical logic. It means that human mind uses this logic too, i.e. it is not an artifact being plain theory far from the human thinking. The fuzzy logic can be proved to include the classical logic. Hence, it covers two sides of human thinking:

- the rigorous thinking being modelled by two-valued logic,
- the vagueness (to a certain extent, of course).

The most important logical rule is modus ponens. E.g. we state

(1) IF the snow falls down near the factory THEN it goes black

Next, we have regarded that the snow really has fallen down near the factory. Then we infer using (1) that it will go black after some time. We have used modus ponens in this inference.

There is no problem from the point of the classical logic. Inspecting the statementes in more details we find, however, that they are vague very much. What does it mean "the snow falls down", "near the factory", "black" ? How much snowfall is enough to be sure that " the snow has fallen down"? Of course, there are many situations in which we cannot have doubts. But there exist borderline cases which are not so clear. Can we be really sure that the given amount of snow is black or it is rather grey? If the notion of "black" had been clear we should not have had to define the absolutely black body in physics.

We have pointed out that vagueness inducess the degrees of truth. In modus ponens, we work with the degree of truth of the antecedent (" the snow has fallen down") and the implication (1) as well. Then the consequent ("the snow will go black") has also the degree of truth which is derived from those of antecedent and implication. It must decrease slightly since from uncertain premises we can derive only more uncertain conclusion. In symbols, the modus ponens can be written down as

$$(2) \qquad \frac{\varphi, \varphi \Rightarrow \psi}{\psi} \left(\frac{\alpha, \beta}{\alpha \otimes \beta}\right)$$

where α, β are the degrees of truth of the antecedent φ and the implication $\varphi \Rightarrow \psi$ respectively and $\alpha \otimes \beta$ is the degree of truth of the consequent ψ where $\otimes$ is the operation of the special multiplication. (cf.[5,6]). The rule (2) concerns elements of vague utterances, i.e., in our case , the concrete snowfall, factory, colour etc. In fuzzy sets theory, we model the meaning of vague notions by fuzzy sets. Hence, the above example of modus ponens is a certain relation among fuzzy sets which is derived from (2).

People use modus ponens very often, sometimes with some modifications. The modifications follow usually from the wealth of natural language. It is possible to use e.g. the modificators (linguistic operators) such as "very", "more or less", "roughly" etc. They slightly change the meaning of the utterance (make it broader or narrower). If e.g. we know that the snow has fallen down very near the factory then from (1) we infer that it will go black very much. Such inferences are not possible to model in the classical logic but they can be modelled in the fuzzy logic. The situation is not, however, so simple and the inference depends heavily on the structure of the natural language. Though some successes have been achieved, much work must still be done in this field in connection with classical linguistics.

Fuzzy logic has practical applications

The above modus ponens rule has inspired fuzzy set theoretists and engineers to many applications. The condition (1) can also be understood to define a non-exact functional dependence. In fuzzy sets theory, it is modelled as a special kind of fuzzy relation. If we denote it by R and the antecedent by A, then the consequent C is given by

$$(3) \qquad C = A \circ R$$

where " $\circ$ " denotes the composition of fuzzy relations. The rule (3) is called the compositional rule of inference and it is the basis of various applications of fuzzy sets. The most successful application is fuzzy controller. It allows us to control ill described processes, or the processes whose description is so complicated that it is not usable in practice. Such processe are e.g. steering the train, steering the river ship, the traffic control at the junction, clinker production in the cement kiln, sawn wood drying etc. All these and other processes have been examited to control using fuzzy controller. They use the rules of the form "if the system is in the state S then do C ". For example, "if the temperature is low and decreasing, then increase the fuel rate", "if the sawn wood is too wet, then slightly decrease the drying line speed" etc. the first commercial application of the fuzzy controller was developed by F.L.Smidth& Co. of Copenhagen for cement kilns . In Japan, the automatic operation of a train was experienced. There are also applications of fuzzy logic in complex system modelling, in expert systems, pattern recognition and in other fields.

Fuzzy logic is a theoretical frame of the fuzzy sets theory

The fundamental notion in the fuzzy sets theory is the notion of membership. A fuzzy set is a collection of some elements, e.g. the collection of all the trees, red colours, living people etc. The crucial fact is that we are not generaly able to decide unambiguously about an element if it belongs to the fuzzy set given or not. If A is a fuzzy set of matches in the given matchbox then for any x we are able to decide if the formula $x \in A$ (x is an element of A) is true or not (i.e. if it is a match from the matchbox or not). Hence, this fuzzy set is a classical set. Considering the fuzzy set B of red colours, we are able only to determine the degree of truth of $x \in B$. We conclude that the membership degree equals to the degree of truth of the predicate $x \in B$. Therefore, fuzzy sets can really be understood as generalization of the notion of the set. The language of the fuzzy sets theory should be the many-valued fuzzy logic.

Any mathematical theory uses symbols which are interpreted in a sense, i.e. we define what we understand by them. We say that the language of the theory has its syntax and semantics. The question now arises, what is the balance between syntax and semantics? In fuzzy logic it proceeds as follows: we make some assumptions (with some degrees of truth) and then we manipulate formally with them. The truth degree of the result is also derived. Now, we ask what is the relation between this truth value and the truth value of the result when being interpreted. If the first one is greater than the second one then our formal reasoning is ill founded. In that case we could derive very nice results with big truth value having, however, a very small truth value in any interpretation. Shortly- we could infere rubbish.

To avoid it , we must have "good" rules of inference at our disposal assuring that the truth value α of the derived result is smaller than or equal to the truth value β of the worst interpretation of it . It can be proved that there exist "good" rules of inference (the notion "good" rule of inference is exactly defined in fuzzy logic such that $\alpha = \beta$. It is a very strong result- the so called completeness theorem - which is a generalization of the famous Gödel's completeness theorem of the classical predicate calculus. The language of the fuzzy sets theory is therefore well founded.

All the operations in the fuzzy sets theory are based on the connectives from fuzzy logic. The basic connectives in fuzzy logic are disjunction ($\vee$), conjunction ($\wedge$), bold conjunction ($\&$) and implication ($\Rightarrow$). Using implication we can define the negation of the formula φ as $\varphi \Rightarrow$ false. If A,B are two fuzzy sets then the predicates $x \in A$ and $x \in B$ can be joined using any of the above connectives, e.g.

$$x \in A \vee x \in B$$

which is the interpretation of union of the fuzzy sets A,B. Similary for the other connectives. A very important fact is that we are able to introduce additional connectives (they must fulfil the so called fitting condition which is not very strong).Similarly, basic quantifiers are existential ($\exists$) and general ($\forall$) ones but additional quantifiers can also be introduced. We obtain a very rich language which seems to be proper to serve as a basis for the modelling of the natural language semantics as wellas for modelling of human thinking in various situations.

Linguistic logic

In our consideration until now, the formula was assumed to attain only single truth value being e.g. the number in the interval $\langle 0,1 \rangle$. What happens if we are able only to express the truth value vaguely in the natural language, i.e. to say e.g. that the value is "high, average" etc? This rather complicated situation can be viewed as a special kind of logic which is called linguistic logic (we can call it also ultrafuzzy logic). Its truth values are fuzzy sets. We may use linguistic expressions e.g. "true, very true,rather false" etc. but some linguistic problems could be met since it may sometimes lead to improper formulations (e.g."roughly true or not very true" and similar).

There exists a principle called the extension principle using which the extensions of connectives in many-valued fuzzy logic can be introduced as connectives in linguistic logic. We construct truth tables with linguistic truth values, formulate rules of inference etc. The situation is, however, very complicated since the operations with fuzzy sets do not form a closed structure. It means that when obtaining the truth value in the form of fuzzy set, we must search the linguistic expression which corresponds to the fuzzy set. This procedure is called linguistic approximation and it is one of the most complicated questions in fuzzy sets theory.

Linguistic fuzzy logic belongs to the boldest departures from classical logic. We do not know if the human mind does work with linguistic truth values directly or they are only the consequence of using natural language when talking about truth. This question must be answered in the future.

Can the fuzzy logic replace the classical logic ?

This question is somewhat provocative. We have mentioned in our previous discussion that the classical logic can be viewed as a special case of the fuzzy logic. It is a very important fact having several consequences:

- classical logic can be studied from various points of view which may help to understand it still more deeply.
- the basic logical laws that are widely accepted are not excluded but extended. Therefore, the exactness of the classical logic having been critised can be avoided if necessary.

There are, however, many situations in which exactness is unavoidable. We cannot refuse it at all. The classical logic will always be the metalanguage of mathematical thinking, it will be unavoidable in many sciences such as physics, cybernetics, electronic and others. Refusing classical logic would lead to anarchy in thinking which is undesirable. Our aim is not negation of the classical logic but its extension to be able to comprehend vagueness and step farther to understanding the natural language.

References

[1] Bellman, R.E., Zadeh, L.A.: Local and fuzzy logics. In: Modern Uses of Multiple-Valued Logic (J.M. Dunn, G. Epstein, eds.). Reidel Publ., Dordrecht 1977.

[2] Gaines, B.R.: Foundations of fuzzy reasoning. Int. J.Man-Machine Studies, 8 (1976), 623-668.

[3] Haack, S.: Do we need "fuzzy logic"? Int. J.Man-Machine Studies 11 (1979), 437-445.

[4] Mamdani, E.H., Gaines, B.R. (eds.): Fuzzy Reasoning and its Application. Academic Press, London 1981.

[5] Novák, V.: First-Order Fuzzy Logic. Submitted to Studia Logica.

[6] Pavelka, J.: On fuzzy logic. Zeitschr. f. math. Logik und Grundlagen d. Math. 25 (1979), 45-52; 119-134; 447-464.

[7] Russel, B.: Vagueness. Australian J. Phil. 1 (1923), 84-92.

[8] Zadeh, L.A.: Making computers think like people. IEEE Spectrum, August 1984, 26-32.

[9] Zadeh, L.A.: The concept of a linguistic variable and its application to approximate reasoning. Information Sciences 8 (1975), 199-249; 301-357; 9 (1975), 43-80.

FIRST-ORDER FUZZY LOGIC AND ITS APPLICATIONS IN FUZZY MODELLING

Vilém Novák [+]

1. Introduction

This paper is a brief survey of the many-valued first-order fuzzy logic in which deep results have been obtained. The most important is the completeness theorem stating that a formula φ is provable in the degree α iff (if and only if) it is true in the degree α, i.e. the syntax and semantics of the first-order fuzzy logic are in the balance. The set of truth values is confined to be either finite chain or the interval of reals $\langle 0,1\rangle$. The careful reasoning on the possible, operations on these sets has shown that the only plausible structure of them (if the completeness property is not to be violated) is that of residuated lattice. This structure is in accordance with experimental results,linguistic analysis and our intuitin. Thus, the first-order fuzzy logic is well-founded and it is a direct generalization of the classical predicate calculus which we obtain when using the Boolean lattice $\{0,1\}$ for the set of truth values. For detailed presentation of the first-order fuzzy logic see [4].

2. The set of truth values

The set of truth values forms a complete, infinitely distributive and residuated lattice, i.e. the structure.

(1) $$\mathcal{L} = \langle L, \vee, \wedge, \otimes, \rightarrow, 1, 0\rangle$$

being the complete lattice and having moreover the properties:

a) For every $\alpha \in L$, $K \subseteq L$

$$\alpha \wedge \bigvee_{\beta \in K} = \bigvee_{\beta \in K} (\alpha \wedge \beta)$$

$$\alpha \vee \bigwedge_{\beta \in K} = \bigwedge_{\beta \in K} (\alpha \wedge \beta)$$

holds.

b) 1,0 are the greatest and the smallest elements respectively and $\otimes$, $\rightarrow$ are additional binary oprations of multiplication and residuation respectively having the properties:

+ Czechoslovak Academy of Sciences, Mining Institute, Hladnovská 7, 710 00 Ostrava 2, Czechoslovakia.

ba) $\langle L, \otimes, 1\rangle$ is a commutative monoid.

bb) For every $\alpha, \beta, \gamma \in L$ such that $\alpha \leq \beta$ it holds

$$\alpha \otimes \delta \leq \beta \otimes \delta$$

$$\alpha \rightarrow \delta \geq \beta \rightarrow \delta$$

$$\delta \rightarrow \alpha \leq \delta \rightarrow \beta$$

$$\alpha \otimes \beta \leq \gamma \text{ iff } \alpha \leq \beta \rightarrow \gamma$$

If $L = \langle 0,1\rangle$ then we define

$$(2) \qquad \begin{aligned} \alpha \otimes \beta &= 0 \vee (\alpha + \beta - 1) \\ \alpha \rightarrow \beta &= 1 \wedge (1 - \alpha + \beta) \end{aligned}$$

for every $\alpha, \beta \in \langle 0,1\rangle$. Similarly, for $L = \{0 = \alpha_0 \leq \ldots \leq \alpha_m = 1\}$ being a finite chain we define

$$(3) \qquad \begin{aligned} \alpha_k \otimes \alpha_p &= \alpha_{\max(0,k+p-m)} \\ \alpha_k \rightarrow \alpha_p &= \alpha_{\min(m,m-k+p)} \end{aligned}$$

where $0 \leq k,p \leq m$, The choice of operations (2), (3) follows from the following facts (cf. [5]):

- If the operation $\rightarrow$ is not conntionous on $\langle 0,1\rangle^2$ then it is not possible to construct fuzzy logic with complete syntax.
- Every residuated lattice (1) with $L = \langle 0,1\rangle$ and such that the operation $\rightarrow$ is continuons $\langle 0,1\rangle^2$ is isomorphic to the above residuated lattice with the operations (2).

We will frequently use the symbols

$$\alpha \leftrightarrow \beta = (\alpha \rightarrow \beta) \wedge (\beta \rightarrow \alpha)$$

$$\neg \alpha = \alpha \rightarrow 0.$$

We enrich the lattice (1) by the set of n-ary operations $\{o_j; j \in Jop\}$ on L fulfilling the <u>fitting condition</u>

$$(4) \qquad (\alpha_1 \leftrightarrow \beta_1)^{k_1} \otimes \ldots \otimes (\alpha_n \leftrightarrow \beta_n)^{k_n} \leq o_j(\alpha_1, \ldots, \alpha_n) \leftrightarrow o_j(\beta_1, \ldots, \beta_n)$$

for every $j \in Jop$ and some $k_1, \ldots, k_n$ where the exponentation is taken in the sense of $\otimes$, and by the set of generalized operations $\{Q_j; j \in Jq\}$

$$Q_j : \mathcal{P}(L) \rightarrow L$$

where $\mathcal{P}(L)$ is the power set of L. We assume $Q_j, j \in Jq$ are <u>regular</u>

generalized operations, i.e. they fulfil the conditions (5) - (9):

(5) $$Q\alpha = \alpha$$

(6) $$\underset{\alpha\in K}{Q}(\alpha \otimes \beta) \leqq (\underset{\alpha\in K}{Q}\alpha) \otimes \beta$$

(7) $$\neg \underset{\alpha\in K}{Q} \neg(\alpha \otimes \beta) \leqq (\neg \underset{\alpha\in K}{Q} \neg\alpha) \otimes \beta$$

(8) $$\bigwedge_{\alpha\in K} \alpha \leqq \underset{\alpha\in K}{Q}\alpha \leqq \bigvee_{\alpha\in K}\alpha \quad , \quad K\neq\emptyset$$

If $K, I \subseteq L$ and $\alpha \leqq \beta$ holds for every $\alpha \in K$ and $\beta \in I$ then

(9) $$\underset{\alpha\in K}{Q}\alpha \leqq \underset{\beta\in I}{Q}\beta$$

The operations $\bigvee, \bigwedge$ are regular generalized operations.

3. Syntax and semantics

The language of the first-order fuzzy logic consists of variables, constants, functional and predicate symbols, binary connectives $\vee, \wedge, \&, \Rightarrow$ (disjunction, conjunction, bold conjuction, implication, a set of n-ary connectives $\{\varrho_j; j \in Jop\}$, symbols for general $\forall$ and existential $\exists$ quantifiers and a set of generalized quantifiers $\{Q_j;\ j\in Jq\}$.

The common recurent definition of terms and formulas can be given. Note that there are introduced also special atomic formulas $\underline{\alpha}$ for every $\alpha \in L$.

A structure for the language of the first-order fuzzy logic is

$$\mathcal{D} = \langle D,\ p_{\mathcal{D}}, \ldots,\ (g_{\mathcal{D}}\ ,\ G), \ldots$$

where D is a set, $p_{\mathcal{D}} \subseteq D^n$ are n-ary fuzzy relations and $g : G^n \rightsquigarrow G$ are n-ary fuzzy functions with the domain $G \subseteq D$ i.e.the functions $g_{\mathcal{D}} : D^n \rightarrow D$ fulfilling the condition

(10) $$\underset{x_1,\ldots,x_n \in D}{G g_{\mathcal{D}}}(x_1,\ldots,x_n) \geqq Gx_1 \wedge \ldots \wedge Gx_n$$

Each n-ary functional symbol is adjoined an n-ary fuzzy function and each n-ary predicate symbol is adjoined an n-ary relation. The interpretation of terms are fuzzy singletons of the form

$$\{{}^{\alpha}/d\}$$

where $d \in D$ and $\alpha \in L$. We also write d_{α} instead of $\{{}^{\alpha}/d\}$ and $d \in_{\alpha} D$ instead of $d_{\alpha} \lessapprox D$.

The interpretion of the formula χ is the truth valuation $\mathcal{D}(\chi) \in L$ defined as follows:

a) $\mathcal{D}(\underline{\alpha}) = \alpha, \quad \alpha \in L.$

b) $\mathcal{D}(a=b) = \begin{cases} 1, & \text{if } \mathcal{D}(a) = \mathcal{D}(b), \text{ i.e. if } \mathcal{D}(a)=d_{\alpha}, \mathcal{D}(b)=d'_{\beta} \\ & \text{then } \alpha=\beta \text{ and } d=d' \\ 0 & \text{otherwise} \end{cases}$

where a,b are terms.

c) $\mathcal{D}(\varphi \vee \psi) = \mathcal{D}(\varphi) \vee \mathcal{D}(\psi)$

d) $\mathcal{D}(\varphi \wedge \psi) = \mathcal{D}(\varphi) \wedge \mathcal{D}(\psi)$

e) $\mathcal{D}(\varphi \,\&\, \psi) = \mathcal{D}(\varphi) \otimes \mathcal{D}(\psi)$

f) $\mathcal{D}(\varphi \Rightarrow \psi) = \mathcal{D}(\varphi) \rightarrow \mathcal{D}(\psi)$

g) $\mathcal{D}(\underline{o}_j(\varphi_1, \ldots, \varphi_n)) = o_j(\mathcal{D}(\varphi_1), \ldots, \mathcal{D}(\varphi_n)), \quad j \in Jop$

h) $\mathcal{D}((\exists x)\varphi) = \bigvee_{d \in_{\alpha} D} \mathcal{D}(\varphi_x[\underline{d}_{\alpha}])$

where $\underline{d}_{\alpha}$ is the name (constant) of the fuzzy singleton d_{α}.

i) $\mathcal{D}((\forall x)\varphi) = \bigwedge_{d \in_{\alpha} D} \mathcal{D}(\varphi_x[\underline{d}_{\alpha}])$

j) $\mathcal{D}((\underline{Q}_j\, x)\varphi) = Q_{j\, d \in_{\alpha} D}(\varphi_x[\underline{d}_{\alpha}])$

where $\varphi_x[\underline{d}_{\alpha}]$ is the formula resulting from φ when substituting the name $\underline{d}_{\alpha}$ for all the free occurences of the variable x in φ.

It can be proved that the definition of the interpretion of formulas behaves well with respect to the substition of terms.

Given the formula $\varphi(x_1, \ldots, x_n)$. It is α-true in the structure $\mathcal{D}$, $\mathcal{D} \models_{\alpha} \varphi$, if

$$\alpha = \bigwedge_{\substack{d_i \in_{\alpha_i} D \\ i=1,\ldots,n}} \mathcal{D}(\varphi_{x_1,\ldots,x_n}[\underline{d}_{1\alpha_1}, \ldots, \underline{d}_{n\alpha_n}])$$

i.e. its truth is given by the infinum over all the truth values of φ in $\mathfrak{D}$ when substituting all the possible names of fuzzy singletons $d_{i \alpha_i}$ in D. The infinum of all the truth values of the formula φ taken over all the structures $\mathfrak{D}$ for the given language is its degree of validity. (is an α-tautology). There are many formulas that can be proved to be tautologies in the degree 1 .

We need also formal rules of inference using which we are able to derive syntactically another formulas. The rule of inference is a pair $r = \langle r^{syn}, r^{sem} \rangle$ where r^{syn} is its syntactical part and r^{sem} its semantical part. We say that an n-ary rule of inference r is sound if

$$\mathfrak{D}(r^{syn}(\varphi_1, \ldots, \varphi_n)) \geqq r^{sem}(\mathfrak{D}(\varphi_1), \ldots, \mathfrak{D}(\varphi_n))$$

holds i any structure $\mathfrak{D}$ for all the formulas φ . The following rules of inference are sound:

Modus ponens

$$r_{MP} : \frac{\varphi, \varphi \Rightarrow \psi}{\psi} \quad \left(\frac{\alpha, \beta}{\alpha \otimes \beta}\right)$$

Generalization

$$r_G : \frac{\varphi}{(\underline{Q}x)\varphi} \quad \left(\frac{\alpha}{\alpha}\right)$$

Generalization of the consequent

$$r_{G,C} : \frac{\varphi \Rightarrow \psi}{\varphi \Rightarrow (\underline{Q}x)} \quad \left(\frac{\alpha}{\alpha}\right)$$

on the assumption that x is not free in φ .

Distributivity of the quantifier

$$r_{DQ} : \frac{(\underline{Q}x) \quad (\varphi \& \psi)}{(\underline{Q}x)\varphi \& \psi} \left(\frac{\alpha}{\alpha}\right)$$

on the assumption that x is not free in ψ. The quantifiers $\underline{Q}$ are supposed to be regular.

4. Fuzzy theories of the first order

The <u>theory</u> $\mathcal{T}$ in the language of the first-order fuzzy logic (fuzzy) theory) is the three-tuple

$$\mathcal{T} = \langle A_L, A_S, \mathcal{R} \rangle$$

where A_L, A_S are fuzzy sets of <u>logical</u> and <u>special axioms</u> respectively and $\mathcal{R}$ is the set of sound rules of inference. The structure $\mathcal{D}$ is a <u>model</u> of the theory $\mathcal{T}$ if

$$A_S \varphi \leqq \mathcal{D}(\varphi)$$

holds for every formula φ. A formula φ is α- true in the theory $\mathcal{T}$, $\mathcal{T} \models_\alpha \varphi$, if

$$\alpha = \bigwedge \{\mathcal{D}(\varphi); \mathcal{D} \text{ is a model of the theory } \mathcal{T}\}.$$

The sequence of formulas

$$\varphi_0, \varphi_1, \ldots, \varphi_n = \varphi$$

is a <u>proof</u> of the formula φ if every φ_i is a special or logical axiom, or it is syntactically derived using some sound rule of inference, The proof has a value given by the grades of membership of the formulas φ_i in A_L and A_S, and by semantic parts r^{sem} of the rules of inference . The formula φ is α- <u>theorem</u> (α- provable) of the theory $\mathcal{T}$, $\mathcal{T} \vdash_\alpha \varphi$, if

$$\alpha = \bigvee \{\beta; \beta \text{ is the value of the proof of } \varphi \text{ in } \mathcal{T}\}.$$

<u>Theorem</u> (validity theorem).

$$\text{If } \mathcal{T} \vdash_\alpha \varphi \text{ then } \mathcal{T} \models_\beta \varphi$$

where $\alpha \leqq \beta$.

The element

$$\beta = \bigvee \{\beta'; \ \beta' = \alpha_1 \otimes \alpha_2 \text{ and } \mathcal{T} \vdash_{\alpha_1} \varphi, \ \mathcal{T} \vdash_{\alpha_2} \urcorner\varphi \text{ for all the formulas } \varphi\}$$

is called the <u>degree of contradictoriness</u> of the theory $\mathcal{T}$. The theory $\mathcal{T}$ is consistent if $\beta = 0$.

The theory $\mathcal{T}$ is <u>complete</u> if it is consistent and $\mathcal{T} \vdash_\alpha \varphi$ follows $\mathcal{T} \vdash_{\urcorner\alpha} \urcorner\varphi$.

<u>Theorem .</u> Every consistent theory $\mathcal{T}$ can be extended to a complete theory .

Theorem. Let $\varphi(x_1,\dots,x_n)$ be a formula of theory $\mathcal{T}$. Then

$$\mathcal{T} \vdash_\alpha \varphi(x_1,\dots,x_n) \quad \text{iff} \quad \mathcal{T} \vdash_\alpha (\forall x_1)\dots(\forall x_n)\, \varphi(x_1,\dots,x_n)$$

Similarly as in the classical predicate calculus, the canomical model of the theory $\mathcal{T}$ can be constructed. Then the following generalizations of the Gödel's completeness theorems can be proved.

Theorem. A theory $\mathcal{T}$ is consistent iff it has a model. A consistent theory has a model $\mathcal{D}_0$ in which

$$\mathcal{T} \vdash_\alpha \varphi \quad \text{iff} \quad \mathcal{D}_0 \models_\alpha \varphi$$

holds for every formula φ.

Theorem. A formula φ is α-theorem of the theory $\mathcal{T}$ iff it is true in the degree α in it:

$$\mathcal{T} \vdash_\alpha \varphi \quad \text{iff} \quad \mathcal{T} \models_\alpha \varphi \, .$$

5. Discussion and applications

The first-order fuzzy logic is a frame or, in certain sense, the language of the fuzzy sets theory. Using it, most results in the fuzzy sets theory and applications can be formulated. The important facts are the following:

- first-order fuzzy logic is complete which ensures the formal reasoning to be semantically founded.
- the system of fuzzy logic allows us to use additional connectives and quantifiers except for the basic ones to be able to develop our models closer to the reality. If the condition (4) is fulfilled and the quantifiers are regular then completeness property is saved.

We will now very briefly demonstrate the way how fuzzy logic could be used..

a) Fuzzy ordering

Fuzzy ordering was introduced in [9]. It is a fuzzy theory in the language $\{\leq\}$ with the following fuzzy set of special axioms:

$$\{{}^1/(\forall x)\,(x \leq x),\ {}^1/(\forall x)(\forall y)\,(\forall z)\,(x\leq y \wedge y \leq z \Rightarrow x \leq z),$$

$$ {}^1/(\forall x)(\forall y)((x\leq y \wedge y\leq x) \Rightarrow x=y)\}$$

The interpretation is a structure

$$\mathcal{D} = \langle D,\ R_\leq \rangle$$

where $R_\leq \subseteq D\text{x}D$ is a binary fuzzy relation with the properties being induced by the special axioms: For every $x \in_\alpha D, y \in_\beta D, z \in_\gamma D$ where $\alpha, \beta, \gamma \in L$

$$R_{\leqq}\langle x,x\rangle = 1$$

$$R_{\leqq}\langle x,z\rangle \geqq R_{\leqq}\langle x,y\rangle \wedge R_{\leqq}\langle y,z\rangle, \quad \text{i.e. } R_{\leqq}\langle x,z\rangle \geqq \bigvee_{y\in D}(R_{\leqq}\langle x,y\rangle \wedge R_{\leqq}\langle y,z\rangle)$$

$R_{\leqq}\langle x,y\rangle \wedge R_{\leqq}\langle y,x\rangle > 0$ follows $x=y$
holds. Considering only fuzzy singletons with membership 1 we obtain fuzzy ordering exactly in the sense of [9] .

b) Fuzzy equality in the sense of A.Pultr [6] .

It is a fuzzy theory in the language $\{\doteq\}$ with the following fuzzy set of special axioms

$$\{{}^1/(\forall x)(x \doteq x),\quad {}^1/(\forall x)(\forall y)(x \doteq y \Leftrightarrow y \doteq x),$$

$$ {}^1/(\forall x)(\forall y)(\forall z)((x \doteq y \,\&\, y \doteq z) \Rightarrow x \doteq z)$$

The interpretation is a structure

$$\mathcal{D} = \langle D, R_{\doteq}\rangle$$

where $R_{\doteq} \subsetneqq D \times D$ is a binary fuzzy relation with the properties being induced by the special axioms: For every $x \in_\alpha D, y \in_\alpha D$, $z \in_\gamma D$ where $\alpha, \beta, \gamma \in L$

$$R_{\doteq}\langle x,x\rangle = 1$$

$$R_{\doteq}\langle x,y\rangle = R_{\doteq}\langle y,x\rangle$$

$$R_{\doteq}\langle x,z\rangle \geqq R_{\doteq}\langle x,y\rangle \otimes R_{\doteq}\langle y,z\rangle$$

Note that replacing the connective & by $\wedge$ in the axiom of transitivity we obtain the similarity relation defined by L.A. Zadeh in [9] .

In [2], M. Katz defined the proximity relation $\sim$ as a fuzzy relation understanding the degrees of truth in an opposite way, i.e. 0 as truth and 1 as false. Moreover, he uses the implication connective with the interpretation ($L = \langle 0,1\rangle$)

$$\alpha \rightarrow \beta = 0 \vee (\beta - \alpha)$$

He presents the similar fuzzy set of special axioms as A.Pultr does with the third one of the form $1/(\forall x)(\forall y)(\forall z)(x\sim y \rightarrow (y\sim z \rightarrow x\sim z))$. Realizing that $\varphi \rightarrow \psi \Leftrightarrow \varphi \,\&\, \neg\psi$ and $x\sim y \Leftrightarrow \neg\, x \doteq y$ are tautologies in the degree 1, it can be easily proved that Katz's proximity theory is equivalent to the Pultr's fuzzy equality (they both have the same models).

c) The heap paradox (Sorites)

This paradox is based on the deliberation that adding a stone to a small heap of stones we get again a small heap of stones. Hence, every heap is small.

In the frame of fuzzy logic, it can be solved e.g. as follows. Let $\varphi_n(x)$ be a formula stating that x having n elements is a small heap. Then the heap fuzzy theory has the following fuzzy set of special axioms $(L=\langle 0,1\rangle)$:

$$\left\{ {}^{1}/(\forall x)\ \varphi_0(x),\ \left\{ {}^{1-\varepsilon}/(\forall x)\varphi_n(x) \Rightarrow (\forall x)\ \varphi_{n+1}(x);\ n=0,1,\ldots,\ \varepsilon>0 \right\}\right\}$$

Our reasoning can be modelled in the form of the proof

$$(\forall x)\ \varphi_0(x)\ [1],\ (\forall x\ \varphi_0(x) \Rightarrow (\forall x)\ \varphi_1\ [1-\varepsilon], (\forall x)\ \varphi_1(x)\quad [1-\varepsilon; r_{MP}], \ldots,$$

$$(\forall x)\varphi_n(x)\quad [0 \vee (1-n\cdot\varepsilon\ );\ r_{MP}\ ]$$

In the square brackets there is given the truth value of the formula and possibly, the rule of inference using which it was derived. We can see that the truth value of the formula $(\forall x)\varphi_n(x)$ decreases with the increasion of n and for $n \geq \frac{1}{\varepsilon}$ the truth that every heap is small is equal to zero.
The ancient paradox appers if we consider the truth of the formula $(\forall x)\varphi_n(x) \Rightarrow (\forall x)\ \varphi_{n+1}(x)$ equal to 1.
This model is linear. We can, of course, use non-linear one if we change the fuzzy set of special axioms.

The reader may see that most papers fuzzifying some part of mathematics, e.g. fuzzy groups[7], lattices,topology and others can be formulated as an axiomatically given first order fuzzy theory. The first -order fuzzy logic has also other applications.Defining new additional connectives and quantifiers, we can model various sides of the natural language semantics and human thinking. Since such procedure does not violate the completeness theorem, our model is well founded.

There is much work still to be done. We do not e.g. know how open fuzzy theories behave. Also the quantifiers are to be investigated in more details. All the operations used must fulfil the fitting condition which e.g. in ⟨0,1⟩ means that the operations must be continuous. It would be very interesting to investigate the behaviour of the theory when the fitting condition is violated in some points. These and many other questions must be answered in the future.

References

[1] Goguen,J.A.: The Logic of Inexact Concepts.Synthese 19 (1968), 325-373

[2] Katz,M.: Controlled-Error Theories of Proximity and Dominance.In: H.J.Skala,S.Termini, E.Trillas eds. Aspects of Vagueness (D.Reidel,Dordecht 1984).

[3] Negoita ,C.V., Ralescu,D.A.: Applications of Fuzzy Sets to Systems Analysis. Birkhäuser-Verlag, Bazel-Stuttgart 1975.

[4] Novák, V.: First-Order Fuzzy Logic. Submitted to Studia Logica.

[5] Pavelka, J.: On fuzzy logic. Zeitschr.f. math. Logik und Grundlagen d.Math. 25(1979),45-52; 119-134; 447-464.

[6] Pultr, A.: Fuzziness and fuzzy equality. In: H.J.Skala,S.Termini, E.Trillas eds. Aspects of Vagueness (D.Reidel,Dordecht 1984).

[7] Rosenfeld,A.: Fuzzy groups. J.Math. Anal. Applic. 35(1971),512-517.

[8] Schoenfield,J.R.: Mathematical Logic. Addison-Wesley,New York 1967.

[9] Zadeh, L.A.: Similarity relations and fuzzy orderings. Inf.Sci 3 (1971), 177-200.

ON THE SOLUTION OF DIRECT AND INVERSE PROBLEMS FOR FUZZY EQUATION SYSTEMS WITH TOLERANCES

M. Wagenknecht, K. Hartmann [1]

1. Introduction

In the present paper we will deal with the following problems:

a) Direct problems

Let us consider

$$\underline{B}_i \leq R_i \circ X \leq \overline{B}_i \quad ; \; i=1,\ldots,N \; . \tag{1}$$

$\underline{B}_i$, $\overline{B}_i$ are given fuzzy sets over a basic space $\mathcal{Y}$. The R_i are fuzzy relations over $\mathcal{X} \times \mathcal{Y}$ whereby $\mathcal{X}$ is another basic space. We are seeking for a fuzzy set X (over $\mathcal{X}$) fulfilling (1). Furthermore we suppose that there are given fuzzy sets B_i over $\mathcal{Y}$ with $\underline{B}_i \leq B_i \leq \overline{B}_i$. " $\circ$ " means a convenient rule of composition. Hereforth we denote this problem by (P_D). We may interpret (P_D) as a generalization of the following problem Find X fulfilling

$$R_i \circ X = B_i \quad ; \quad i=1,\ldots,N \; . \tag{2}$$

We denote this problem by (P_{DE}). It originates among others from the fuzzification of classical nonlinear equation systems. The functions occuring there have been replaced by fuzzy relations R_i, the right-hand sides by B_i and the unknowns by X. The situation that we investigate (P_D) instead of (P_{DE}) is partially reasoned by the well-known fact that (P_{DE}) is not always solvable (cf. /3/ and /4/). There are several possibilities to generalize (P_{DE}) and to define a new conception of solvability, respectively. We refer to Pedrycz /3/ where a solution of (P_{DE}) is determined in a numerical environment, and to Gottwald where a solvability index is used for characterization of the solution set. (See also the present volume).

The above stated problem (P_D) has two advantages in our opinion: On one hand, the demand to the user to determine the tolerances $\underline{B}_i$, $\overline{B}_i$ seems to be fairly transparent - he may predict where tolerances (and in which quantities) are acceptable for him. On the other hand, we are able to describe the struture of the solution set of (P_D) completely in the case of finite $\mathcal{X}$, $\mathcal{Y}$ whereby the corresponding algorithms can be realized in an easy manner. The limitation to finite basic spaces does not seem to be too restrictive, since many practical problems can be grasped in this way. However, some of the achieved results are also valid for compact $\mathcal{X}$, $\mathcal{Y}$ (e.g. in Euclidian space), when the correspon-ding membership-functions are upper semi-continuous. Nevertheless, the numerical handling of the rule of composition (to mention only one

1) Technische Hochschule "Carl Schorlemmer" Leuna-Merseburg
DDR - 4200 Merseburg, Otto-Nuschke-Straße

item) will be rather complicated so that we restrict ourselves to finite sets.

b) Inverse problems

(P_I^1) : Let be given B_i, R_i and upper tolerances $\overline{B}_i$. How to determine the lower tolerances $\underline{B}_i$ such that the problem (P_D) becomes consistent (there is at least one solution X)?

(P_I^2) : See (P_I^1), but this time the lower tolerances $\underline{B}_i$ are given.

Both problems are of interest from the standpoint of application: On one hand, we are able to evaluate the "amount of consistency" if we are able to solve (P_I^1) and (P_I^2) in a certain "optimal" manner (the greater the necessary tolerances the more contradictions we have in our problem). On the other hand, it may be interesting to know much about the tolerances given by practical demands, especially the difference to those necessary for consistency.

For the solution of (P_D), (P_I^1) and (P_I^2) we particularly use a method described by Higashi and Klir in /2/ for the solution of $(P_{D\blacksquare})$ in the case N=1. In section 2 we deal with a generalization of this method to our problems. Section 3 contains some more basic definitions from fuzzy set theory. The next two sections are devoted to the solution of the problems under consideration. Numerical examples are given to enlight the situation. In the 6th section we summarize and give an outlook as well. We should emphasize that all facts are given without proofs on account of shortage of room. A complete paper is available from the authors.

2. Some facts from the theory of partially ordered sets (posets)

In the sequel we assume P to be a poset with partial order "$\leq$".

Definition 2.1. Let $\check{p} \in P$. We call it a minimal element for P iff for arbitrary $p \in P$ we have $p=\check{p}$, whenever $p \leq \check{p}$.

Definition 2.2. We call $\hat{p} \in P$ the greatest element of P iff for all $p \in P$ we have $p \leq \hat{p}$.

Henceforth we denote the set of all minimal elements of P by P_o. Then we can state the following lemmata:

Lemma 2.1. Let P be finite. Then we have $P_o \neq \emptyset$ and for all $p \in P$ we can find $\check{p} \in P_o$ with $\check{p} \leq p$.

Lemma 2.2 (see /2/). Let $P' \subseteq P$. If there is for all $p \in P$ a $p' \in P'$ with $p' \leq p$ then we have $P_o = P'_o$.

We observe that the inclusion $P_o \subseteq P'$ follows from the above assumptions and it should not enter into the formulation of the lemma (as stated in /2/).

Now we define a point-to-set mapping F as follows:

1. $F : P \to \mathcal{P}(P)$ with $\mathcal{P}(P)$ as the power set of P.
2. Let $p \in P$ and $q \in F(p)$. Then $q \leq p$ holds.
3. Let $p_1, p_2 \in P$ and $p_1 \leq p_2$. Then we have $F(p_1) \subseteq F(p_2)$.
4. For all $p \in P$ the set $F(p)$ has at least one minimal element.

The following theorems have been proved in /2/ for (P_{DE}) and N=1. Here we shall give generalizations useful for our objective.

Theorem 2.1. We have $p \in P_o$ iff $F(p)=p$.

Theorem 2.2. Let $p \in P$. Then there exists $p' \in P_o$ with $p' \leq p$.

Theorem 2.3. Suppose $\hat{p}$ to be the greatest element of P. Then the following inclusion holds: $P_o \subseteq F(\hat{p}) \subseteq P$.

Theorem 2.4. Let $\hat{p}$ be the greatest element of P. Then we have $F_o(\hat{p})=P_o$.

The last statement enlights the usefulness of the mapping F: If $\hat{p}$ and $F_o(\hat{p})$ are easy to abtain (in comparison with a straight forward determination of P_o) the above method may be advantageously.

3. Basic definitions from fuzzy set theory

Let $\mathcal{X} = \{ \mathscr{x}_1, \ldots, \mathscr{x}_m \}$, $\mathcal{Y} = \{ y_1, \ldots, y_n \}$. By $\mathcal{F}(\mathcal{X})$, $\mathcal{F}(\mathcal{Y})$, $\mathcal{F}(\mathcal{X} \times \mathcal{Y})$ we denote the sets of all fuzzy sets over $\mathcal{X}$, $\mathcal{Y}$, $\mathcal{X} \times \mathcal{Y}$ (Cartesian product). Let $X \in \mathcal{F}(\mathcal{X})$. By μ_X we denote the corresponding membership-function. For sake of lucidity we will also use the following abbrevations: $\mu_X(\mathscr{x}_i)=x_i$, $\mu_Y(y_j)=y_j$, $\mu_{R_k}(\mathscr{x}_i, y_j)=r^k_{ij}$.

Definition 3.1(Rule of composition). Let $h : [0,1] \times [0,1] \to [0,1]$ be a continuous function of two arguments with following properties:

1. $h(x,y)$ is non-decreasing in both arguments;
2. $h(x,h(y,z))=h(h(x,y),z)$ - associativity;
3. $h(x,y)=h(y,x)$ - commutativity;
4. $h(x,1)=x$; $h(x,0)=0$;

for all $x,y,z \in [0,1]$.

Now let $X \in \mathcal{F}(\mathcal{X})$, $R \in \mathcal{F}(\mathcal{X} \times \mathcal{Y})$. Then we define $Y=R \circ X \in \mathcal{F}(\mathcal{Y})$ by $y_j = \max_{1 \leq i \leq m} h(x_i, r_{ij})$; $j=1,\ldots,m$. We notice that h is a special kind of so-called t-norms,playing an important role in fuzzy set theory. For a detailed analysis we refer to /1/ and /3/.

Definition 3.2. Let X_1, X_2 be two fuzzy sets over the same basic space. We say that $X_1 \subseteq X_2$ iff $\mu_{X_1} \leq \mu_{X_2}$ for all elements of the basic space.

Definition 3.3(α-operation). Let $a,b \in [0,1]$. Then we put

$$a \alpha b = \begin{cases} 1 & \text{for } a \leq b \,; \\ \max\limits_{z \in h^{-1}(a,b)} z & \text{for } a > b \,; \end{cases}$$

with $h^{-1}(a,b) = \{ z \in [0,1] : h(a,z) = b \}$.
Sanchez was the first who used α-operations (among others) to resolve fuzzy relation equations (see also /4/).

<u>Definition 3.4.</u> Let $Y \in \mathcal{F}(\mathcal{Y})$, $R \in \mathcal{F}(\mathcal{X} \times \mathcal{Y})$. Then we put $X = R \alpha Y$ by $x_i = \min\limits_{1 \leq j \leq n} r_{ij} \alpha y_j$; $i = 1, \ldots, m$.

By the help of Definitions 3.1 and 3.2 we are now able to give an exact mathematical formulation of our problems (P_D), (P_I^1) and (P_I^2).

4. <u>The direct problem (P_D)</u>

Now we are going to investigate (P_D). By $\mathcal{V}$ we denote the set of all solutions of (P_D). Furtheron we suppose $\mathcal{F}(\mathcal{X})$, $\mathcal{F}(\mathcal{Y})$ and $\mathcal{F}(\mathcal{X} \times \mathcal{Y})$ to be partially ordered by inclusion according to Definition 3.2. Next we will give a simple criterion for $\mathcal{V} \neq \emptyset$. As a first step we state

<u>Theorem 4.1.</u> Put $X_i^{max} = R_i \alpha \bar{B}_i$; $i = 1, \ldots, N$. Then

$$X^{max} = \bigwedge_{i=1}^{N} X_i^{max} \tag{3}$$

represents the greatest fuzzy set fulfilling

$$R_i \circ X \subseteq \bar{B}_i \; ; \; i = 1, \ldots, N \; ; \tag{4}$$

(here "$\wedge$" denotes the min-operator being applied componentwise).

Now we can give the following important

<u>Theorem 4.2.</u> We have $\mathcal{V} \neq \emptyset$ iff $X^{max} \in \mathcal{V}$. In this case X^{max} is the greatest element of $\mathcal{V}$.

We will deal now with the construction of a mapping F as defined in section 2. For simplification we set $\mu_{\underline{B}_i}(y_k) = \underline{b}_k^i$ and similarly for $\bar{B}_i$. Let $X \in \mathcal{F}(\mathcal{X})$. We define

$$I_k^i(X) = \{ j \in \{1, \ldots, m\} : h(x_j, r_{jk}^i) \geq \underline{b}_k^i \} \; ; \tag{5}$$

$$I^i(X) = \mathop{\times}_{k=1}^{n} I_k^i(X) \; ; \; i = 1, \ldots, N \; ;$$

(our denotation is connected with that in /2/). We choose $f^i \in I^i(X)$ and construct the following mapping: $[f^i(X)] : \mathcal{V} \to \mathcal{F}(\mathcal{X})$ with

$$[f^i(X)]_j = \begin{cases} \max\limits_{k \in J_{f^i}(j)} \; \min\limits_{z \in h^{-1}(r_{jk}^i, \underline{b}_k^i)} z & \text{for } J_{f^i}(j) \neq \emptyset \,; \\ 0 & \text{for } J_{f^i}(j) = \emptyset \,; \end{cases} \tag{6}$$

with j running from 1 to m.

The index set $J_{f^i}(j)$ is defined as follows: Let $f^i=(f^i(1),\dots,f^i(n))$. Then $J_{f^i}(j)=\{l\in\{1,\dots,n\} : f^i(l)=j\}$. Finally we set $f=\overset{N}{\underset{i=1}{\times}} f^i\in I(X)$ with $I=\overset{N}{\underset{i=1}{\times}} I^i(X)$ and define a mapping $f(X)\colon \mathcal{V}\to\mathcal{F}(\mathcal{X})$ by

$$[f(X)]_j=\max_{1\le i\le N}[f^i(X)]_j \; . \tag{7}$$

Now we can give the desired definition of F:

$$F(X)=\{[f(X)] : f\in I(X)\} \; . \tag{8}$$

Theorem 4.3. The following statements are true:

1. For $X\in\mathcal{V}$ we have $F(X)\subseteq\mathcal{V}$.
2. Let $f\in I(X)$. Then $[f(X)]\le X$.
3. Let $X,Z\in\mathcal{V}$ with $X\le Z$. If $[f(X)]\in F(X)$ then we obtain $[f(Z)]\in F(Z)$ and $[f(X)]=[f(Z)]$.

This theorem shows that the above constructed mapping F satifies the conditions of the second section. The fourth condition (existence of minimal elements of F(X)) is trivially fulfilled, since F(X) is a finite set and therefore Lemma 2.1 holds. As a consequence we get:

Theorem 4.4. The sets $\mathcal{V}$ and $F(X^{max})$ have the same minimal elements, i.e. $\mathcal{V}_0=F_0(X^{max})$.

With respect to the structure of $\mathcal{V}$ we formulate

Theorem 4.5(Representation Theorem). The following equality is valid:

$$\mathcal{V}=\bigcup_{\check{X}\in\mathcal{V}_0}[\check{X}, X^{max}], \text{ whereby } [\check{X}, X^{max}]=\{X\in\mathcal{F}(\mathcal{X}) : \check{X}\le X\le X^{max}\}.$$

("$\cup$" means classical set union).

Hence we may interpret $\mathcal{V}$ as a union of a finite number of "fuzzy intervals". From the Representation Theorem we deduce a simple criterion for unique solvability of (P_D):

Theorem 4.6. $\mathcal{V}$ consists of only one element iff $X^{max}=F(X^{max})$.

From the theory given up to now we can derive the following algorithm for the construction of $\mathcal{V}_0$:

1. Computation of X^{max} and test whether $X^{max}\in\mathcal{V}$ or not. In the latter case we have $\mathcal{V}=\emptyset$, that is, inconsistency. STOP! Otherwise go to 2.
2. Construction of $I^i(X^{max})$ for $i=1,\dots,N$.
3. Determination of $F(X^{max})$.
4. Computation of $F_0(X^{max})$ for instance by pairwise comparison.

Despite the principal feasibility of the algorithm the computation of $F(X^{max})$ can be rather extensive. That is why we will give a theorem allowing a further reduction of computational effort. For this end we introduce yet the following denotations:

$$F^i(X)=\{[f^i(X)] : f^i\in I^i(X)\}\; ; \tag{9}$$

$$\check{F}(X)= \{ Z \in \mathcal{F}(\mathcal{X}): Z= \bigvee_{i=1}^{N} X^i , X^i \in F_0^i(X) \}. \tag{10}$$

We have of course $Z \in F(X)$ iff $Z= \bigvee_{i=1}^{N} X^i$ with $X^i \in F^i(X)$; ("$\vee$" denotes the max-operator being applied componentwise). Now we can state

Theorem 4.7. The sets $F(X)$ and $\check{F}(X)$ have the same minimal elements for $X \in \mathcal{V}$, that is

$$F_0(X) = \check{F}_0(X). \tag{11}$$

Example. Before listing up numerical data we give useful remarks. If we take $h(x,y)=\min(x,y)$ we get

$$a \alpha b = \begin{cases} 1 & \text{for } a \leq b ; \\ b & \text{otherwise.} \end{cases} \tag{12}$$

Hence we obtain

$$\left[f^i(X)\right]_j = \begin{cases} \max\limits_{k \in J_f{}^i(j)} b_k^i & \text{for } J_f{}^i(j) \neq \emptyset ; \\ 0 & \text{otherwise.} \end{cases} \tag{13}$$

Now we take $\mathcal{X}=(x_1, x_2, x_3)$, $\mathcal{Y}=(y_1, y_2, y_3)$. Furtheron let $N=2$ and $R_1 = \begin{pmatrix} .1 & 0 & .6 \\ .5 & .1 & .3 \\ .2 & .9 & 0 \end{pmatrix}$, $R_2 = \begin{pmatrix} .5 & .1 & .2 \\ 1. & .2 & .6 \\ .2 & .8 & .1 \end{pmatrix}$. Moreover we take $h(x,y)= \min(x,y)$. At first we consider

$$R_1 \circ X = B_1 , \quad R_2 \circ X = B_2 \tag{14}$$

with $B_1=(.3,.2,.4)$ and $B_2=(.5,.1,.3)$. We notice that (14) is inconsistent: The first equation has the unique solution $X^*=(.4,.3,.2)$ which does not fulfil the second equation. Now we provide (14) with the following tolerances: $\underline{B}_1=(.1,.1,.2)$, $\overline{B}_1=(.5,.3,.6)$; $\underline{B}_2=(.3,0,.1)$, (15) $\overline{B}_2=(.7,.2,.5)$. Applying Theorem 4.2 we obtain $\mathcal{V} \neq \emptyset$ with $X^{max}=(1.,.5,.2)$. Therefore we get $I_1^1(X^{max})=\{1,2,3\}$; $I_2^1=\{2,3\}$; $I_3^1=\{1,2\}$; $I_1^2=\{1,2\}$; $I_2^2=\{1,2,3\}$; $I_3^2=\{1,2,3\}$. Omitting elementary computations we get $F_0^1=\{(.2,.1,0); (.1,0,.1); (0,.2,0)\}$, $F_0^2=\{(.3,0,0); (0,.3,0)\}$ which leads to $\mathcal{V}_0=F_0=\check{F}_0=\{(.3,.1,0); (.3,0,.1); (0,.3,0)\} =\{\check{X}_1,\check{X}_2,\check{X}_3\}$ and therefore $\mathcal{V}= \bigcup_{i=1}^{3} [\check{X}_i , X^{max}]$ using Theorem 4.5.

For the case $\underline{B}_1=B_1=\overline{B}_1$ ($\underline{B}_2$, $\overline{B}_2$ as before) we get $X^{max}=(.4,.3,.2)$ and of course $X^{max}=X^*$. When formally using our algorithm we obtain: $I_1^1=\{2\}$; $I_2^1=\{3\}$; $I_3^1=\{1\}$; $I_1^2=\{1\}$; $I_2^2=\{1,2,3\}$; $I_3^2=\{1,2,3\}$ and therefore $F_0^1=\{(.4,.3,.2)\}$; $F_0^2=\{(.3,0,0)\}$ and hence $\check{F}=\{(.4,.3,.2)\}$. But then $\mathcal{V}_0= \check{F}_0=X^{max}$, that is, $\mathcal{V}=X^{max}$ as already stated above.

5. The inverse problems (P_I^1) and (P_I^2)

5.1. The problem (P_I^1)

Before investigating the structure of the solution set of the problem (the formulation of which has been given in the introduction) we again shall give necessary denotations. So we define

$$\mathcal{F}^N(\mathcal{Y}) = \{(Y_1,\dots,Y_N) : Y_j \in \mathcal{F}(\mathcal{Y});\ j=1,\dots,N\}. \tag{16}$$

We assume this set to be a poset by componentwise ordering: With $\vec{Y}^1, \vec{Y}^2 \in \mathcal{F}^N(\mathcal{Y})$ we define $\vec{Y}^1 \leq \vec{Y}^2$ iff $Y_j^1 \leq Y_j^2$; $j=1,\dots,N$ when using the notation $\vec{Y}^i=(Y_1^i,\dots,Y_N^i)$; $i=1,2$. Now we are going to define the set of solutions of (P_I^1):

$$\underline{\mathcal{B}}(\vec{B}, \vec{B}) = \{\vec{\underline{B}} \in \mathcal{F}^N(\mathcal{Y}) : (1) \text{ is consistent for given } \vec{B}, \vec{B} \in \mathcal{F}^N(\mathcal{Y}) \text{ with } \vec{\underline{B}} \leq \vec{B} \leq \vec{B}\}. \tag{17}$$

We have of course $\underline{\mathcal{B}} \subseteq \mathcal{F}^N(\mathcal{Y})$. The following statement gives us the possibility to determine the extremal elements of $\underline{\mathcal{B}}$:

<u>Theorem 5.1.</u> The set $\underline{\mathcal{B}}$ has a greatest element $\vec{\underline{B}}^{max}$ which can be computed in the following way:

$$\underline{B}_i^{max} = B_i \wedge R_i \circ X^{max}\ ;\ i=1,\dots,N \tag{18}$$

whereby X^{max} is the fuzzy set given by Theorem 4.1. We will end this section with the remark that $\underline{\mathcal{B}}$ always possesses a least element, namely the empty fuzzy set.

5.2. The problem (P_I^2)

Now we deal with the characterization of the set of feasible upper tolerances when the lower ones are given. Again we start with definitions: For a fuzzy set under consideration we denote the maximal one (over the corresponding basic space) by $\mathbb{E}$. So $\vec{\mathbb{E}} \in \mathcal{F}^N(\mathcal{Y})$ means $\mu_{\mathbb{E}_i}(y_j)=1$; $i=1,\dots,N$; $j=1,\dots,n$; etc. Now we define the counterpart of $\underline{\mathcal{B}}$ from the last section:

$$\overline{\mathcal{B}}(\vec{\underline{B}}, \vec{B}) = \{\vec{\overline{B}} \in \mathcal{F}^N(\mathcal{Y}) : (1) \text{ is consistent for given } \vec{\underline{B}}, \vec{B} \in \mathcal{F}^N(\mathcal{Y}) \text{ with } \vec{\underline{B}} \leq \vec{B} \leq \vec{\overline{B}}\}. \tag{19}$$

While in the previous section we could state that $\underline{\mathcal{B}}$ is never empty it may happen that $\overline{\mathcal{B}}$ is. Namely, let us choose $\vec{B}=\vec{\mathbb{E}} \in \mathcal{F}^N(\mathcal{Y})$. Then we compute the corresponding $\vec{\underline{B}}^{max}$ according to the above theory. It is clear that for $\vec{\underline{B}} \nleq \vec{\underline{B}}^{max}$ the set $\overline{\mathcal{B}}$ will be empty. Nevertheless, we have the following simple proposition: If $\overline{\mathcal{B}} \neq \emptyset$ then $\vec{B}^{max}$ (as defined above) is the greatest element (and vice versa).

Now we are going to take up the determination of minimal elements of $\overline{\mathcal{B}}$, that is $\mathcal{B}_0$. Next we consider the following problem: Find $X \in \mathcal{F}(\mathcal{X})$ with

$$\underline{B}_i \leq R_i \circ X \leq \vec{B}_i^{max} \; (=\mathbf{B}) \; ; \; i=1,\ldots,N \; . \tag{20}$$

We denote the solution set of this problem by $\mathcal{V}'$. According to Theorem 4.2 we know that $\mathcal{V}' \neq \emptyset$ iff $X'^{max} \in \mathcal{V}'$ with $X'^{max} = \mathbf{B} \in \mathcal{F}(\mathcal{X})$. From Theorem 4.4 we have $\mathcal{V}'_0 = F_0(\mathbf{B})$. Let

$$\check{\bar{\mathcal{B}}} = \{ \vec{\bar{B}} \in \mathcal{F}^N(\mathcal{Y}) : \bar{B}_i = R_i \circ \check{X} \vee B_i ; \; \check{X} \in \mathcal{V}'_0 ; \; i=1,\ldots,N \} . \tag{21}$$

Then we can state

Theorem 5.2. The sets $\check{\bar{\mathcal{B}}}$ and $\bar{\mathcal{B}}$ have the same minimal elements, i.e. $\check{\bar{\mathcal{B}}}_0 = \bar{\mathcal{B}}_0$

Example. First we consider (P_I^1) with the data of the above example, whereby $\bar{B}_1$ and $\bar{B}_2$ are taken from (15). We will determine the maximal lower tolerances according to (18). We have already obtained that $X^{max} = (1.,.5,.2)$. Furtheron, $R_1 \circ X^{max} = (.5,.2,.6)$; $R_2 \circ X^{max} = (.5,.2,.5)$. Hence we get $\underline{B}_1^{max} = (.3,.2,.4)$; $\underline{B}_2^{max} = (.5,.1,.3)$. This shows that the lower tolerances from (15) are not "optimal" in the sense of feasibility.

For illustration of (P_I^2) we choose the lower tolerances from (15) and determine the set $\bar{\mathcal{B}}_0$ due to Theorem 5.2. After corresponding computations we obtain $\mathcal{V}'_0 = \mathcal{V}_0$ with $\mathcal{V}_0$ as in section 4. From this we have $\bar{\mathcal{B}}_0 = (B_1, B_2)$, that means, the right-hand sides of (14) are the minimal upper tolerances in this example.

6. Concluding remarks

In the present paper we investigated the structure of solutions of a certain class of fuzzy equation systems. It turns out that the solution set can be described in the terms of poset theory. The applied method can be extended to problems of fuzzy modelling with tolerances (i.e., we look for a fuzzy relation when fuzzy inputs and outputs are given). This will be matter of further investigation. Besides we have interesting connections to fuzzy eigen set problems in given regions as it will be shown in future.

References

/1/ Gottwald, S.: Characterizations of the Solvability of Fuzzy Equations (to be published).

/2/ Higashi, M. and Klir, G.J.: Resolution of Finite Fuzzy Relation Equations. Fuzzy Sets and Systems 13 (1984), 65-82.

/3/ Pedrycz, W.: Fuzzy Control and Systems. Preprint of Dept. of Math., Delft Univ. of Technology, Delft 1982.

/4/ Sanchez, E.: Resolution of Composite Fuzzy Relation Equations. Inf. Control 30 (1976), 38-48.

ON SOME THEORETICAL PROBLEMS CONCERNING THE CONSTRUCTION OF FUZZY CONTROLLERS

Siegfried Gottwald[1)]

Abstract The fuzzy controllers, a very simple case of expert systems, form an useful tool e. g. in control engineering. Starting with case studies for laboratory devices in the seventieth, now there exist already microelectronical realizations e.g. for cement industry to be purchased. Nevertheless, there are truly difficult theoretical problems connected with the construction of such fuzzy controllers, especially with the question of the existence of controllers really realizing given intuitive demands. We state here some generalized existence criteria, and we discuss a modification procedure which allows to transform a system of control rules that admits only an approximately correct realization into a system which is correctly realizable.

1. Introduction

Fuzzy controllers form a - simple, but very useful - special case of expert systems. They are of great importance for processing fuzzy information. And they potentially are suited for doing this in an automatized manner.

Therefore, in control engineering the fuzzy controllers are forming a topic of extensive studies. Essentially there are case studies that present applications or applicational ideas for various fields of automatic control, viz. in systems controlled manually by a human being acting as operator of the process under control. Also applications in so-called "soft" science are worth to be mentioned, i. e. applications in such fields where human beings play a central role. We mention only (i) a model of organizational behaviour of WENSTOP (1976) where linguistic rules are used in describing the organizational behaviour at a mining company, (ii) a model of social-psychological power theory of MULDER that proceeds in the framework of fuzzy sets, cf. KICKERT (1978), (iii) the use of fuzzy sets in medical diagnosis, cf. e.g. ADLASSNIG/KOLARZ (1982), and (iv) the fuzzy risk analyzer as presented in SCHMUCKER (1984).

The essential way in which fuzzy sets enter those fields of research is the use of suitable linguistic variables. This ist obvious in the "soft" cases, but the same happens with technical systems: while speaking about control strategies to be applied we too are able to recognize fuzzy concepts - partly by reference to some human being(s) acting in the control process, partly by the fact that the goal of the control is not precisely defined or definable or it is - practically - impossible to formulate it transparantly in a nonfuzzy manner.

Generally, one starts from the idea that a fuzzy controller is extracted from some linguistic protocol through some kind of formaliza-

1) Karl-Marx-Universität, Sektion M.-l. Philosophie, WB Logik, DDR-7010 Leipzig

tion and determination of fuzzy sets which correspond to the notions in the protocol. This means to start the considerations from a given system of control rules

$$A_i \Longrightarrow B_i , \qquad i = 1, \ldots, N. \qquad (1)$$

These control rules connect the values of some input variable(s) with the values of some output variable(s) of the control process. Assume that the input and the output variables are fixed in advance, and that all the control rules of the system (1) too are given in advance and refer to the same variables. Hence, if m input variables and n output variables are fixed in some order then each A_i is to be taken as a m-tupel, and each B_i as a n-tupel of fuzzy subsets of suitable universes of discourse.

Of course, all the m values of the input variables put together in each A_i may be combined more closely by taking their fuzzy cartesian product, i.e., one may change from a "many input" fuzzy controller to a "single input" one. Theoretically this transformation is possible in every case, but practically is may be more convenient to choose another policy - to split such a "many input" fuzzy controller into some controllers with fewer inputs each. For the industrial plant (a combination of a steam engine and a boiler) considered in MAMDANI/ASSILIAN (1975) one has originally a 4-input 2-output fuzzy controller which splits in a natural way into two 2-input 1-output controllers, presented as heater algorithm and as throttle algorithm in that paper. And only those two 2-input fuzzy controllers should be transformed into 1-input controllers by the method mentioned before. Thus we consider here for the sake of simplicity only the case of 1-put controllers. Also we restrict attention to the 1-output fuzzy controllers despite of the problems which may occur if the real output of a (transformed) 1-input fuzzy controller does not have the form of a fuzzy cartesian product of fuzzy sets. It may be expected that in such a critical case some interaction exists between the different output variables; to treat such a case to be theoretically an open problem.

2. Methods to realize fuzzy controllers

At present there are mainly three versions how to realize a system (1) of control rules as an algorithm capable of automated performance:

(i) the original method of MAMDANI/ASSILIAN (1975),

(ii) the use of a fuzzy discretization and the direct activation of the control rules as indicated in HOLMBLAD/ØSTERGAARD (1982),

(iii) the transformation of the system (1) of control rules into a system of fuzzy equations and the solution of the latter system.

In every case, a control rule $A \Longrightarrow B$ is interpreted as an IF-THEN statement which has to be modelized with fuzzy sets.

In version (i) the idea is to convert each control rule into a fuzzy relation and finally to "collect" all those particular fuzzy relations into one final fuzzy relation that constitutes the fuzzy controller corresponding to system (1). The most simple realization here is to take

$$A_i \times B_i \qquad \text{for} \qquad A_i \rightarrow B_i \,, \tag{2}$$

i.e. the fuzzy cartesian product of input and output value of control rule i instead of control rule i. The fuzzy cartesian product of two fuzzy sets A,B is as usual defined through

$$\mu_{A \times B}(x,y) = \mu_A(x) \wedge \mu_B(y), \tag{3}$$

$\wedge$ for minimum. The final fuzzy relation R constituting the fuzzy controller for system (1) then is defined as

$$R = \bigcup_{i=1}^{N} (A_i \times B_i) \tag{4}$$

with membership degree in this "big union" defined as supremum of the membership degrees in all the fuzzy relations $A_i \times B_i$, i.e.

$$\mu_R(x,y) = \bigvee_{i=1}^{N} (\mu_{A_i}(x) \wedge \mu_{B_i}(y)) \tag{5}$$

with $\bigvee$ for supremum.

Of course, the construction with $\wedge$ in (3) as well as that with $\bigvee$ in (5) may be changed. Sometimes the product or also other t-norms have been considered in (3) instead of $\wedge$ as well as suitable generalizations of the usual implication; cf. e.g. MIZUMOTO (1982) for an extended list of proposals to code fuzzy IF-THEN statements by fuzzy relations.

Having given the fuzzy controller as in (4) the output fuzzy set B for any input fuzzy set A is determined as the full image of A under R which by abuse of notation often is written as the relational product of the fuzzy set A with the fuzzy relation R as

$$B = A \circ R = R''A. \tag{6}$$

The membership degrees are

$$\mu_{R''A}(y) = \bigvee_{x \in \underline{U}} (\mu_A(x) \wedge \mu_R(x,y)); \tag{7}$$

here $\underline{U}$ denotes the universe of discourse for the input fuzzy sets of system (1).

For version (ii) to realize a fuzzy controller corresponding to a system (1) of control rules we assume that the input fuzzy sets A_i of the rules of (1) form a suitable covering of the universe of discourse $\underline{U}$. In this case we say that the input fuzzy sets $A_1, \ldots, A_N$ form a fuzzy discretization of this universe $\underline{U}$. Having given such a fuzzy discretization we avoid a direct construction of a fuzzy relation as in (4) which represents the fuzzy controller, instead for any input fuzzy set A we directly calculate the output fuzzy set B from the system (1) of control rules. To do this we first determine for each control rule i the value

$$\beta_i^A = \mathrm{Poss}(A/A_i) = \bigvee_{x \in \underline{U}} (\mu_A(x) \wedge \mu_{A_i}(x)) \tag{8}$$

and with these values we take

$$B = \bigcup_{i=1}^{N} \beta_i^A \cdot B_i \quad . \tag{9}$$

which means for the membership degrees to take

$$\mu_B(y) = \bigvee_{i=1}^{N} \beta_i^A \cdot \mu_{B_i}(y) \ . \tag{10}$$

Again, of course, here the choice of the product in formulas (9), (10) for the modification of the output fuzzy sets B_i with the values from (8) is a point for discussion. Other t-norms as e.g. the minimum $\wedge$ or the algebraic product § defined as

$$p \ \S \ q = \max (0, p+q-1) \tag{11}$$

may so as well. The intuitive idea realized this way simply is the following one: the fuzzy output B for the fuzzy input A is composed out of the output B_i of the control rules of (1) according to the "degrees" (8) to which the i-th control rule is "activated" by the input A.

It is interesting to note that the choice of minimum instead of product in formula (9) immediately gives

$$\begin{aligned} \mu_B(y) &= \bigvee_{i=1}^{N} (\mu_{B_i}(y) \wedge \bigvee_{x \in \underline{U}} (\mu_A(x) \wedge \mu_{A_i}(x))) \\ &= \bigvee_{x \in \underline{U}} (\mu_A(x) \wedge \bigvee_{i=1}^{N} (\mu_{A_i}(x) \wedge \mu_{B_i}(y))) \\ &= \mu_{R''A}(y) \end{aligned} \tag{12}$$

for the fuzzy relation R as defined in (4), (5). Therefore, the approach of MAMDANI/ASSILIAN (1975), i.e. version (i) here, in this case is subsumed under the version (ii).

Version (iii) accepts the construction of the fuzzy output of the fuzzy controller R for system (1) in case of any fuzzy input A according to formula (6) resp. (7). The difference with version (i) consist in the fact that now there is no direct and simple construction of the fuzzy controller R as in (2), (4), instead the system (1) of control rules is transformed into a system

$$B_i = R''A_i, \qquad i = 1, \ldots, N \tag{13}$$

of fuzzy equations with R as an unknown fuzzy relation which has to be determined. This means, system (13) has to be solved.

Unfortunately, for practical purposes till now it is quite difficult to follow this third way. The reason is that effectively calculable methods are lacking for the determination of solutions of system (13), On the other side the great advantage of version (iii) is to give - if available - a solution of system (13) whereas version (i) as well as version (ii) may give no solution of system (13) also in cases this system is solvable.

For finite universes of discourse for the input and the output fuzzy sets in (1) CZOGALA/DREWNIAK/PEDRYCZ (1982) give an algorithm to determine the lower solutions of o n e fuzzy equation of type $B = R"A$ and characterize all solutions in terms of the lower ones. Additionally it is known that the fuzzy intersection of all upper, i.e. greatest solutions of the equations $B_i = R"A_i$ is the greatest solution of system (13) in case that system is solvable - the proof given in GOTTWALD (1984) for another type of equations works for the present case too. Here we will look for solvability criteria for systems (13) of fuzzy equations. The criteria we get will not only give a yes-no-answer but provide us with a degree of solvability of systems of type (13). In this way they moreover give information about approximate solvability of such systems of fuzzy equations. Hence, in some sense those criteria evaluate a kind of "quality" of systems of control rules (1). Also other ways to evaluate some "quality" of control systems (1) are possible and sometimes useful, cf. GOTTWALD/PEDRYCZ (1985) for the discussion of some variants.

3. Solvability of systems of fuzzy equations

Since the pioneering paper SANCHEZ (1976) a lot of papers studied the solvability of fuzzy equations. But almost no attention was given to systems of fuzzy equations. It seems that GOTTWALD (1984) is the first source where a solvability criterion for systems of fuzzy equations is given. The type of equations considered there is an arithmetical one as in SANCHEZ (1984), but the main idea works for systems of fuzzy equations of type (13) as well. The result may be stated as:

The system (13) of fuzzy equations has a solution iff the fuzzy relation

$$R_o = \bigcap_{i=1}^{N} (A_i \between B_i) \tag{14}$$

is a solution of system (13); and if (13) has a solution then this fuzzy relation (14) is the greatest one.

For a proof we refer to GOTTWALD (198x). Here the "big intersection" of fuzzy relations in formula (14) is defined via the infinum of the membership degrees in all the fuzzy relations $A_i \between B_i$. And those fuzzy relations are defined as

$$\mu_{A_i \between B_i}(x,y) = \mu_{A_i}(x) \alpha \mu_{B_i}(y) \tag{15}$$

using the α-operation of SANCHEZ (1976) which is given as

$$p \alpha q = \begin{cases} 1, & \text{if } p \leq q \\ q, & \text{otherwise.} \end{cases} \tag{16}$$

Thus we have a criterion which gives us the greatest solution of (13) if there exist any solution at all. But we like to have more. For, we have to expect that the control rules of system (1) will influence one another in (at least some) practically important concrete cases, cf. CZOGALA/PEDRYCZ (1981), GOTTWALD (1984a). And, furthermore, we have to expect that such an interactivity of the control rules of system (1) will cause the nonexistence of solutions of system (13).

Guided by those considerations we look for approximate solutions too. For this we have to have some distance for fuzzy sets. The function defined by

$$d(A,B) = \bigvee_{x\in \underline{U}} |\mu_A(x) - \mu_B(x)| \tag{17}$$

is a suitable candidate for such a distance. Additionally, we consider another such dinstance of fuzzy sets given by

$$d^+(A,B) = \bigvee_{\substack{x\in \underline{U} \\ \mu_A(x) \neq \mu_B(x)}} (1 - (\mu_A(x) \wedge \mu_B(x))). \tag{18}$$

For both distances d and d^+ from (17), (18) there is a deeper theoretical background because both these distances are in a canonical manner connected with fuzzy equality relations, i.e. with many valued generalizations of the usual identity of crisp sets. Such generalized identity relations are definable for evera lower semicontinous t-norm using the fact that for each such t-norm T a residuation φ_T is given as asual in lattice theory by

$$p \;\varphi_T\; q = \sup \{z \mid p \;T\; z \leq q\}, \tag{19}$$

cf. GOTTWALD (198x). And as well as the t-norms are generalizations of the conjunction operation of classical logic, those residuation operators are generalizations of the implication operator. With those analogies in mind we put

$$eq_T(A,B) = \bigwedge_{x\in \underline{U}} (\mu_A(x) \;\varphi_T\; \mu_B(x)) \;\; T \;\; \bigwedge_{x\in \underline{U}} (\mu_B(x) \;\varphi_T\; \mu_A(x)) \tag{20}$$

and have with the value $eq_T(A,B)$ a degree that characterizes to which extend the fuzzy sets A,B are equal. Now we get

$$d(A,B) = 1 - eq_{\S}(A,B) \quad\text{and}\quad d^+(A,B) = 1 - eq_{\wedge}(A,B). \tag{21}$$

As a side remark we mention that by $1-eq_T(A,B)$ a metric is defined as long as the t-norm T is greater than the algebraic product § from (11) and lower semicontinuous, i.e. left continuous; of GOTTWALD (198xx).

To get a characterization of approximate solvability of system (13) of fuzzy equations we consider the following solvability index

$$s^+ = \bigwedge_{R\in F_2(\underline{V})} \bigvee_{i=1}^{N} d^+(R"A_i, B_i) \tag{22}$$

with $F_2(\underline{V})$ for the class of all fuzzy (binary) relations in the universe $\underline{V}$. Here $\underline{V}$ is supposed to be the common universe of discourse for the input and output fuzzy sets of system (1) of control rules. Obviosly, this index s^+ is the infinum of all the maximal distances $d^+(R''A_i, B_i)$ of realized and intended outputs of the fuzzy controller corresponding to (1), i.e. s^+ characterizes a kind of quality of realizability of the system (1). At the same time s^+ characterizes the best possible approximate solution of system (13) of fuzzy equations in sense of the distance d^+. The idea to consider s^+ according to (22) is quite natural also from a logical point of view: $1 - s^+$ is simply the truth value of the sentence "system (13) of fuzzy equations has a solution" formulated in a suitable language of many valued logic, cf. GOTTWALD (198x).

Of course, if system (13) has a solution then $s^+ = 0$. Conversely, for finite input and output universes from $s^+ = 0$ it follows that the system (13) really has a solution.

The most interesting fact now is that this solvability index s^+ can be determined without reference to the class $F_2(\underline{V})$ of fuzzy relations. It is proven in GOTTWALD (198x) that

$$s^+ = \bigvee_{i=1}^{N} d^+(R_o''A_i, B_i) \tag{23}$$

for the fuzzy relation R_o from (14). This fuzzy relation R_o is determined solely by the input and output fuzzy sets of the system (1) of control rules, hence also the solvability index s^+ is already fully determined by the control rules resp. by the known data of the system (13) of fuzzy equations.

If one prefers to consider the distance d instead of d^+ which was used in formula (22) then the corresponding solvability index to be discussed has to be

$$s = \bigwedge_{R \in F_2(\underline{V})} \left(1 - \mathop{\&}_{i=1}^{N} (1 - d(R''A_i, B_i))\right) \tag{24}$$

where $\&$ is the finite iteration of the algebraic product (11) and where now we have to take

$$\mu_{R''A}(y) = \bigvee_{x \in \underline{U}} (\mu_A(x) \ \S \ \mu_R(x,y)), \tag{25}$$

i.e. we too have to substitute $\S$ for $\wedge$ in (7).

Again, $s = 0$ if the system (13) of fuzzy equations - with $R''A_i$ understood as in (25) - has a solution; and for finite universes of discourse for the input and output fuzzy sets conversely $s = 0$ implies the existence of a solution of (13). Only the full characterization (23) of the solvability index is lacking now. We only get bounds. To formulate them we have to consider the residuation operator $\varphi_\S$ which by (19) is given as

$$p \ \varphi_\S \ q = \min(1, 1-p+q), \tag{26}$$

i.e. which is exactly the Łukasiewicz implication operator of many valued logic. This operator then has to be used in (15) instead of α ; and the "big intersection" in (14) has to be built from those modified fuzzy relations (15). Call the resulting fuzzy relation R_1. Then it holds true that

$$s \geq 1 - \mathop{\&}_{i=1}^{N} (1-d(R_1"A_i,B_i)), \tag{27}$$

which is the lower bound. An upper bound we first get in a more complicated form as an upper bound of the N-th iteration of s by §:

$$\mathop{\&}_{i=1}^{N} s \leq 1 - \mathop{\&}_{i=1}^{N} (1 - d(R_1"A_i,B_i)). \tag{28}$$

Because § is monotonically increasing in both arguments so is its iteration &, hence it is not too hard get really an upper bound for s from (28).

4. Solvability index and manipulation of fuzzy data

For concrete applications of fuzzy controllers it may happen that the fuzzy data A_i,B_i which constitute the system (1) of control rules resp. the system (13) of fuzzy equations themselves are given only approximately. In engineering practice this often is the case if the system (1) of control rules is given at the very beginning in linguistic form; e.g. some expert may have fixed his everyday experience, or the control rules concern a field like medical diagnosis with mainly linguistic information on essential aspects of the process under control. In some of those cases it is natural to assume that smaller membership degrees of the fuzzy data are less reliable than higher ones. Hence, let us agree to change the system (1) of control rules in such a way that small membership degrees of the fuzzy data are "free for small changes". Let us try to make such a change with the effect that the modified system of fuzzy equations (13) has a greater solvability index as the original one.

Assume that a "small" threshold level u is fixed. With respect to this threshold level change the fuzzy data "below membership degree u". To do this let us introduce for any fuzzy set A its u-modification $\mathrm{mod}_u A$ by the membership function

$$\mu_{\mathrm{mod}_u A}(x) = \max(\mu_A(x), u). \tag{29}$$

Now we change the fuzzy data in this manner and consider instead of system (13) of fuzzy equations the modified system

$$\mathrm{mod}_u B_i' = R"\mathrm{mod}_u A_i , \qquad i = 1, \ldots, N. \tag{30}$$

Denote by $s^+(u)$ resp. by $s(u)$ the solvability index from (22) resp. (24) of the modified system (30).

Obviously, with $u = 1$ the modified system (30) is (trivial but) solvable. And with $u = 0$ the modified system (30) is the original one

(13). It is proven in GOTTWALD/PEDRYCZ (198 x) that in case of normal input fuzzy sets in (1) it holds true for those solvability indices that

$$0 \leq s^+(u) \leq 1 - u \quad \text{and} \quad 0 \leq s(u) \leq 1 - \mathop{\&}_{i=1}^{N} u \tag{31}$$

and furthermore that

$$u_1 \leq u_2 \implies s^+(u_1) \geq s^+(u_2). \tag{32}$$

It is an open problem to characterize those t-norms for which (32) holds true for the corresponding solvability indices.

For practical purposes it seems especially interesting to consider for given fuzzy data A_i, B_i the threshold level

$$u_o^+ = \inf \{u \mid s^+(u) = 0\} \quad \text{resp.} \quad u_o = \inf\{u \mid s(u)=0\} \tag{33}$$

Those threshold levels give besides the solvability indices s^+ resp. s a second kind of evaluation of the suitability of the system (1) of control rules as a realizable fuzzy model. Furthermore, threshold level (33) can be used for a heuristic procedure to change the original system (1) of control rules into such a system of control rules which constitutes a realizable fuzzy model, i.e. which has a corresponding (13) of fuzzy equations which is solvable. Such a method may proceed in the following steps:

- calculate threshold (33) for a given system (1) of control rules;
- determine a solution s_o of the modified system (30) where the modification is done with $u=u_o$ resp. $u=u_o^+$;
- change the output fuzzy sets in the rules of (1) to $S_o"A_i$.

Of course, the last step can be modified in different ways. The essential point with this step is to use the solution S_o to move from system (1) of control rules to a suitably modified system with solvable system of corresponding fuzzy equations. Suitably modified here means that the "new" fuzzy data differ from the "old" ones only in membership degrees below threshold (33).

5. Solvability index and tolerance solutions

In discussing the solvability of system of fuzzy equations WAGENKNECHT/HARTMANN (1985) develop the idea to do this within some given tolerance. For the system (13) of fuzzy equations a very simple variant of those ideas is to look for a fuzzy relation R such that

$$B_i - t_l \subseteq R"A_i \subseteq B_i + t_u, \quad i = 1, \ldots, N \tag{34}$$

where $\subseteq$ is the usual inclusion of fuzzy sets, the real numbers t_l, t_u are the lower and the upper tolerance, and the fuzzy sets $B_i - t_l$, $B_i + t_u$ are defined by their membership fuctions as

$$\mu_{B_i - t_l}(x) = \max(0, \mu_{B_i}(x) - t_l), \tag{35}$$

$$\mu_{B_i + t_u}(x) = \min(1, \mu_{B_i}(x) + t_u). \tag{35}$$

Let us reconsider the fuzzy relation R_1 which was used in (27). From GOTTWALD (198x) we get that

$$R_1 " A_i \subseteq B_i \quad \text{for all } i = 1, \ldots, N. \tag{36}$$

Now straightforward calculations with § for the conjunction operator yield for the solvability index s of system (13)

$$\mu_{R_1 " A_i}(x) \geq \mu_{B_i}(x) + s - 1 . \tag{37}$$

Therefore, by (36) the results from GOTTWALD (198x) give that in (34) the upper tolerance may be choosen as $t = 0$; and because of (37) the lower tolerance may correspondingly be choosen as $t_l = 1 - s$ in case of finite universes of discourse for the input and output data of (1) resp. as $t_l = 1 - s - \delta$ for some small positive real δ in the general case.

Acknowledgment

The results reported in this paper have to a considerable extend been realized in common work with Witold Pedrycz of Gliwice Technical University. Moreover, it was in numerous discussions with him that the author realized the value the approach through fuzzy equations has for the consideration of fuzzy controllers.

References

Adlassnig, K.- P., and G. Kolarz (1982). CADIAC-2: Computer-assisted medical diagnosis using fuzzy subsets. In: Approximate Reasoning in Decision Analysis (M M Gupta, E Sanchez eds.), North-Holland, Amsterdam, 219-247.

Czogala, E., J. Drewniak and W. Pedrycz (1982). Fuzzy relation equations on a finite set. Fuzzy Sets Syst. 7, 89-101.

Czogala, E., and W. Pedrycz (1981). Some problems concerning the construction of algorithms of decision-making in fuzzy systems. Intern. J. Man-Machine Stud. 15, 201-211.

Gottwald, S. (1984). On the existence of solutions of systems of fuzzy equations. Fuzzy Sets Syst. 12, 301-302.

Gottwald, S. (1984a). Criteria for non-interactivity of fuzzy logic controller rules. In: Large Scale Systems: Theory and Applications 1983, Proc. 3rd IFAC/IFORS Symp. Warsaw 1983 (A Straszak ed.), Pergamon, Oxford, 69-79.

Gottwald, S. (198x). Characterizations of the solvability of fuzzy equations. Elektron. Informationsverarb. Kybernet. (in print).

Gottwald, S. (198xx). Fuzzy set theory with t-norms and φ-operators. In: Topics on the Mathematics of Fuzzy Systems (A diNola, A G S Ventre eds. - to appear).

Gottwald, A., and W. Pedrycz (1985). Problems of the design of fuzzy controllers. In: Approximate Reasoning in Expert Systems (M M Gupta et al. eds.), North-Holland, Amsterdam (in print).

Gottwald, S., and W. Pedrycz (198x). Solvability of fuzzy relational equations and manipulation of fuzzy data. Fuzzy Sets Syst. (to appear).

Holmblad, L.P., and J.-J. Østerdaard (1982). Control of a cement kiln by fuzzy logic. In: Fuzzy Information and Decision Processes (M.M. Gupta, E Sanchez eds.), North-Holland, Amsterdam, 389-399.
Kickert, W.J.M. (1978). Fuzzy Theories of Decision Making. Martinus Nijhoff, Leiden.
Mamdani, E.H., and S. Assilian (1975). An experiment in linguistic synthesis with a fuzzy logic controller. Intern. J. Man-Machine Stud. 7, 1-13.
Mizumoto, M. (1982). Fuzzy inference using max -$\wedge$ composition in the compositional rule of inference. In: Approximate Reasoning in Decision Analysis (M.M. Gupta, E Sanchez eds.), North-Holland, Amsterdam, 67-76.
Sanchez, E. (1976). Resolution of composite fuzzy relation equations. Inform. and Control 30, 38-48.
Sanchez, E. (1984). Solution of fuzzy equations with extended operations. Fuzzy Sets Syst. 12, 237-248.
Schmucker, K.J. (1984). Fuzzy Sets, Natural Language Computations, and Risk Analysis. Computer Sci. Press, Rockville/Md.
Wagenknecht, M., and K. Hartmann (1985). On the solution of direct and inverse problems for fuzzy equation systems with tolerances. This volume.
Wenstop, F. (1976). Deductive verbal models of organizations. Intern. J. Man-Machine Stud. 8, 293-311.

ESSENTIALS OF FUZZY SETS THEORY: A LINGUISTIC PERSPECTIVE

Mario Fedrizzi[1)] and Janusz Kacprzyk[2)]

Abstract. A brief introduction to fuzzy sets theory is given . A fuzzy set and fuzzy relations are defined. Basic operations on fuzzy sets are given. A linguistic approach, based on fuzzy sets, is presented to provide tools to use linguistic variables, fuzzy conditional statements, etc. for the analysis of systems. Some more important applications in, e.g., control engineering, management science, etc., are reviewed.

1. INTRODUCTION

The purpose of the paper is to provide the reader with a brief introduction to fuzzy sets theory. Emphasis is on the essence and meaning of concepts, operations, etc. Mathematics is used only instrumentally, that is as a tool, and only to a necessary degree. In all our considerations a linguistic perspective is adopted. First, because it is the most illustrative for a practice-oriented user, and, second, virtually all applications of fuzzy sets theory have been strongly based on linguistic interpretations of fuzzy sets. This will certainly best match the purpose of the volume to stimulate interest and applications of fuzzy sets in diverse domains of systems analysis.

To see fuzzy sets in a proper perspective, let us notice that analysis and modeling of any real world phenomenon or process must take into account an inherent uncertainty as to, e.g., options, dynamics, outcomes, etc. Conventionally, this proceeds in some probabilistic or statistical terms. However, in many cases that uncertainty is not due to randomness but to some imprecision. This may be: ambiguity, i.e. the association with a given object of a number of alternative meanings, generality, i.e. the application of the symbol's meaning to a multiplicity of objects, and vagueness, i.e. a lack of clear-cut boundaries of the set of objects to which the symbol (meaning) is applied. Notice that all the above imprecisions, in particular vagueness, may be viewed as some result of natural languages used by humans.

We will consider in the sequel this last type of imprecision, vagueness, to be called fuzziness, and show how fuzzy sets theory may provide mathematical tools to formally represent and manipulate vagueness. Our discussion will strongly rely on linguistic interpretations as they are most illustrative.

We will also sketch some more successful applications of fuzzy sets, mainly in control engineering and management science.

2. BASIC ELEMENTS OF FUZZY SETS THEORY

Existence and importance of vagueness in many real world problem, in particular with relevant human elements, has been recognized for a long time (van Gigch and Pipino [6]). However, its adequate formal treatment was inhibited by the lack of a proper

1) Institute of Statistic and Operations Research, University of Trento, Via Verdi,26 38100 Trento, Italy.
2) Systems Research Istitute, Polish Academy of Sciences, Newelska 6, 01-447 Warsaw, Poland.

mathematical apparatus (Gaines [5]). A breackthrough occured in 1965 when L.A. Zadeh [16] proposed his simple and elegant theory of fuzzy sets. It has gained a tremendous popularity and found applications in virtually all fields.

2.1. The idea of a fuzzy set

The purpose of a (conventional) set may be viewed to characterize a concept. For instance, "integers between 5 and 8" may be uniquely represented by the set $\{5,6,7,8\}$. The set can be equated with its characteristic function $\phi : X \to \{0,1\}$, where $X = \{x\}$ is a universe of discourse; $\phi(x) = 0$ means that x does not belong to the set, while $\phi(x) = 1$ means that x belongs.

To be able to characterize vague concepts, say "numbers more or less equal to 6", which are beyond the scope of the conventional set theory, Zadeh [16] proposed in 1965 the idea of a fuzzy set in which the characteristic function is replaced by the so called membership function

$$\mu : X \to [0,1] \qquad (.1)$$

which associates with each element $x \in X$ its grade of membership $\mu(x) \in [0,1]$. Now, $\mu(x) = 0$ means that x does not belong to the set, $\mu(x) = 1$ means that x fully belongs, and $0 < \mu(x) < 1$ means that x belongs to some extent; a partial membership is therefore possible.

For example, a fuzzy set A ="numbers close to 6" may be represented by $\mu_A(x)$ shown in Fig. 1. Thus, 6 fully belongs to A, and the more a number differs from 6, the lower its grade of membership.

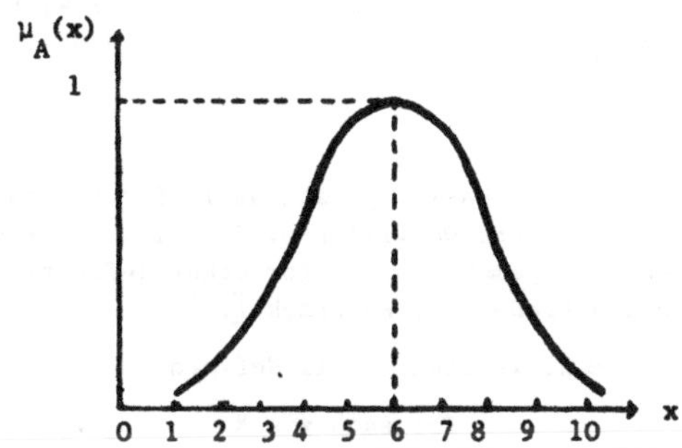

Fig. 1

The transition from belongingness to non-belongingness is now therefore not clear-cut and abrupt as in the conventional set but gradual. Let us also notice that instead of an "objective" characteristic function, we have now a subjective membership function.

Now, formally, a fuzzy set A in X, $A \subseteq X$, is defined as a set of pairs

$$A = \{(\mu_A(x), x)\} \qquad \text{for each } x \in X \qquad (2)$$

In practice, $X = \{x_1, \ldots, x_n\}$ and (2) is written as $A = \mu_A(x_1)/x_1 + \ldots + \mu_A(x_n)/x_n$; " + " is in a set-theoretic sense, and $\mu_A(x)/x$ is the pair $(\mu_A(x), x)$. Moreover, $\mu_A(x)/x$ is omitted if $\mu_A(x) = 0$.

The concept of a fuzzy set makes it possible to develop a consistent theory of fuzzy sets.

2.2. Basic definitions and properties

Two fuzzy sets $A, B \subseteq X$ are equal, written $A = B$, iff (i.e., if and only if)

$$\mu_A(x) = \mu_B(x) \qquad \text{for each } x \in X \tag{3}$$

Example. If $X = \{1,2,3\}$, $A = 0.1/1 + 0.5/2 + 1/3$, $B = 0.2/1 + 0.5/2 + 1/3$, and $C = 0.1/1 + 0.5/2 + 1/3$, then $A = C$ but $A \neq B \neq C$.

Let us notice that the above definition of equality is very rigid (two fuzzy sets can either be or not equal). For a "milder" approach introducing a degree of equality, see Bandler and Kohout [1].

A fuzzy set $A \subseteq X$ is contained in a fuzzy set $B \subseteq X$, written $A \subset B$, iff

$$\mu_A(x) \leq \mu_B(x) \qquad \text{for each } x \in X \tag{4}$$

As above, this is a rigid definition. For a milder one, with a degree of containment, see Bandler and Kohout [1].

A fuzzy set $A \subseteq X$ is normal iff $\mu_A(x) = 1$ for at least one $x \in X$; otherwise it is subnormal.

The support of $A \subseteq X$, written supp A, is the following (nonfuzzy) set

$$\text{supp } A = \{x \in X : \mu_A(x) > 0\} \tag{5}$$

The α-cut (α-Level set) of $A \subseteq X$, written A_α, is $A_\alpha = \{x \in X : \mu_A(x) \geq \alpha\}$.

Example. If $X = \{1,2,3,4,5\}$ and $A = 0.1/1 + 0.5/2 + 0.8/3 + 1/4$, then supp $A = \{1,2,3,4\}$, and $A_{0.1} = \{1,2,3,4\}$, $A_{0.5} = \{2,3,4\}$, $A_{0.8} = \{3,4\}$, and $A_1 = \{4\}$.

2.3. Basic operations on fuzzy sets

As in the conventional set theory, the basic operations in fuzzy sets theory are also the complement, union and intersection. We will provide below their original definitions due to Zadeh [16] which are commonly used; for other definitions, see, e.g., Bellman and Zadeh [2], Dubois and Prade [3], Kacprzyk [10].

The complement of a fuzzy set $A \subseteq X$, written A', is defined as

$$\mu_{A'}(x) = 1 - \mu_A(x) \qquad \text{for each } x \in X \tag{6}$$

The union of $A, B \subseteq X$, written $A+B$, is defined as

$$\mu_{A+B}(x) = \mu_A(x) \vee \mu_B(x) \qquad \text{for each } x \in X \tag{7}$$

where $a \vee b = \max(a,b)$.

The intersection of $A, B \subseteq X$, written $A \cap B$, is defined as

$$\mu_{A \cap B}(x) = \mu_A(x) \wedge \mu_B(x) \qquad \text{for each } x \in X \tag{8}$$

where $a \wedge b = \min(a,b)$.

Example. If $X = \{1,2,3,4\}$, $A = 0.2/1 + 0.5/2 + 0.8/3 + 1/4$, and $B = 1/1 + 0.8/2 + 0.5/3 + 0.2/4$, then:

$A' = 0.8/1 + 0.5/2 + 0.8/3$

$A + B = 1/1 + 0.8/2 + 0.8/3 + 1/4 \qquad A \cap B = 0.2/1 + 0.5/2 + 0.5/3 + 0.2/4$

Linguistically, the complement corresponds to the negation "not", the union to the connective "or", and the intersection to the connective "and".

For some other operations, as well as a discussion of the definitions' adequacy, see, e.g., Kacprzyk [10].

2.4. Fuzzy relations and some related topics

As in the conventional mathematics, a fuzzy relation is also one of key concepts in fuzzy mathematics. Formally, a (binary) fuzzy relation R between the two (non fuzzy) sets X and Y is a fuzzy set in the Cartesian product X × Y, $R \subseteq X \times Y$, and hence

$$R = \{(\mu_R(x,y)/(x,y))\} \qquad \text{for each } (x,y) \in X \times Y \qquad (9)$$

Example. If X = {horse, donkey} and Y = {mule,cow}, then the fuzzy relation R = "similar" may be
"similar" = 0.8/(horse,mule) + 0.4/(horse,cow) + 0.9/(donkey,mule) + 0.2/(donkey,cow) =

$$= \begin{array}{c c} & \begin{array}{cc} \text{mule} & \text{cow} \end{array} \\ \begin{array}{c} \text{horse} \\ \text{donkey} \end{array} & \begin{bmatrix} 0.8 & 0.4 \\ 0.9 & 0.2 \end{bmatrix} \end{array}$$

The matrix representation of a fuzzy relation, as shown above, is very convenient in practice.

Evidently, since a fuzzy relation is a fuzzy set, all the properties, operations, etc. given for fuzzy sets remain valid for fuzzy relations.

The (max-min) composition of two fuzzy relations $R \subseteq X \times Y$ and $S \subseteq Y \times Z$, written RoS is defined as the following fuzzy relation $RoS \subseteq X \times Z$

$$\mu_{RoS}(x,z) = \max_{y \in Y} (\mu_R(x,y) \wedge \mu_S(y,z)) \qquad \text{for each } (x,z) \in X \times Z \qquad (10)$$

Example. For the below R and S, RoS is

$$\underset{x}{\overset{R,\ y}{\begin{array}{c|ccc} & 1 & 2 & 3 \\ \hline 1 & 0.3 & 0.8 & 1 \\ 2 & 0.9 & 0.7 & 0.4 \end{array}}} \quad o \quad \underset{y}{\overset{S,\ z}{\begin{array}{c|cccc} & 1 & 2 & 3 & 4 \\ \hline 1 & 0.7 & 0.6 & 0.4 & 0.1 \\ 2 & 0.4 & 1 & 0.7 & 0.2 \\ 3 & 0.5 & 0.9 & 0.6 & 0.8 \end{array}}} \quad = \quad \underset{x}{\overset{RoS,\ z}{\begin{array}{c|cccc} & 1 & 2 & 3 & 4 \\ \hline 1 & 0.5 & 0.9 & 0.7 & 0.8 \\ 2 & 0.7 & 0.7 & 0.7 & 0.4 \end{array}}}$$

The Cartesian product of two fuzzy sets $A \subseteq X$ and $B \subseteq Y$, written $A \times B$, is defined as

$$\mu_{A \times B}(x,y) = \mu_A(x) \wedge \mu_B(y) \qquad \text{for each } x \in Y,\ y \in Y \qquad (11)$$

Evidently, the above definitions and properties for binary fuzzy relations may be immediately extended for n-ary fuzzy relations.

For a detailed discussion of fuzzy relations, see Kaufmann [11].

2.5. The extension principle and fuzzy numbers

The extension principle, one of the most important and powerful tools in fuzzy sets theory, addresses the following fundamental problem: if there is some relationship between nonfuzzy entities, what is its equivalent between fuzzy quantities? This makes it therefore possible to extend some known models or algorithms involving nonfuzzy variables to the case of fuzzy variables.

Formally, if $A_1 \subseteq X_1, \ldots, A_n \subseteq X_n$ are some fuzzy sets, and $f: X_1 \times \ldots \times X_n \rightarrow Y$ is some (nonfuzzy) function, $y = f(x_1, \ldots, x_n)$, then due to the extension principle the fuzzy set $B \subseteq Y$ induced by the fuzzy sets $A_1,\ldots,A_n$ through the (non fuzzy) function f is given by

$$\mu_B(y) = \max_{(x_1,\ldots,x_n) \in X_1\times\ldots\times X_n \,:\, y = f(x_1,\ldots,x_n)} \bigwedge_{i=1}^{n} \mu_{A_i}(x_i) \qquad (12)$$

where $\bigwedge_{i=1}^{n} a_i = a_1 \wedge \ldots \wedge a_n$.

Example. If $X_1 = \{1,2,3\}$, $X_2 = \{1,2,3,4\}$, f is addition, i.e. $y = x_1 + x_2$, $A_1 = 0.1/1 + 0.6/2 + 1/3$, and $A_2 = 0.6/1 + 1/2 + 0.5/3 + 0.1/4$, then

$$B = A_1 + A_2 = 0.1/2 + 0.6/3 + 0.6/4 + 1/5 + 0.5/6 + 0.1/7$$

Notice that we use here "+" both in the arithmetic and set theoretic sense.

The extension principle can be used for any operation, not only the arithmetic ones as shown above. For details, see Dubois and Prade [3].

However, the use of the extension principle for the arithmetic operations is practically most relevant as it is a prerequisite for the concept of a fuzzy number.

Usually, a fuzzy number is defined as a normal and convex fuzzy set in the real line, $A \subseteq R$. The fuzzy numbers play the same role in fuzzy models as the (real) numbers in non fuzzy models, hence their universal importance.

The four basic arithmetic operations, i.e. addition, subtraction, multiplication and division, between two fuzzy numbers $A,B \subseteq R$ can be obtained by applying the extension principle, respectively:

$$\mu_{A+B}(z) = \max_{x+y=z} (\mu_A(x) \wedge \mu_B(y)) \qquad (13)$$

$$\mu_{A-B}(z) = \max_{x-y=z} (\mu_A(x) \wedge \mu_B(y)) \qquad (14)$$

$$\mu_{AB}(z) = \max_{xy=z} (\mu_A(x) \wedge \mu_B(y)) \qquad (15)$$

$$\mu_{A/B}(z) = \max_{x/y=z:y\neq 0} (\mu_A(x) \wedge \mu_B(y)) \qquad (16)$$

for each $z \in R$, $x, y \in R$. Similarly, we can define other operations. For details on fuzzy numbers and related topics, see Dubois and Prade [3].

3. A FUZZY-SET-BASED LINGUISTIC APPROACH TO THE ANALYSIS OF SYSTEMS

Traditionally, for the analysis of both "hard" (e.g., technological) and "soft" (e.g., social) systems "precise" tools of the convetional mathematics have been used. The success has been often much less than expected. Looking for a reason and remedy, Zadeh [17] has formulated the so-called principle of incompatibility stating that as the complexity of a problem increases, our ability to analyze it in precise yet relevant terms diminishes. A solution may be to use some linguistic descriptions which are commonly used by humans providing a simple but often adequate enough tool to describe even the most complex situation, and an extraordinary information aggregation.

3.1. Zadeh's linguistic approach

The above has led Zadeh [17,18] to propose a new approach to the analysis of complex systems, decision processes, etc., the so called linguistic approach.

Basically, Zadeh's linguistic approach is based on the notion of a linguistic variable. It plays the same role as a (numerical) variable in conventional models but takes on linguistic terms, say "high", "around 5", "a little less than high", as its values. The values of linguistic variables are then equated with appropriate fuzzy sets.

To characterize a relationship (dependence) between linguistic variables, a fuzzy conditional statement is used. For instance, if L and K are linguistic variables taking on fuzzy values $A \subseteq X$ and $B \subseteq Y$, respectively, then a dependence between L and K may be given by a fuzzy conditional statement IF(L=A) THEN (K=B), or shorter IF A THEN B.

It is usually assumed that

$$\text{IF A THEN B} = A \times B \tag{17}$$

i.e. is equated with the Cartesian product $A \times B$ being a fuzzy relation.

And analogously, for more complex fuzzy conditional statements

$$\text{IF A THEN B ELSE C} = A \times B + A' \times C \tag{18}$$

$$\text{IF } A_1 \text{ THEN } B_1 \text{ ELSE IF } A_2 \text{ THEN } B_2 \text{ ELSE...ELSE IF } A_n \text{ THEN } B_n =$$

$$= A_1 \times B_1 + A_2 \times B_2 + \ldots + A_n \times B_n \tag{19}$$

An immediate problem associated with the use of fuzzy conditional statements is: if L takes on a value, what is the value of K implied by the dependence between L and K? The answer gives the compositional rule of inference: if $R \subseteq X \times Y$ is a fuzzy relation representing a dependence between L and K (a fuzzy conditional statement), L is taking on a value A, then the induced value of K is

$$\mu_B(y) = \max_{x \in X} (\mu_A(x) \wedge \mu_R(x,y)) \qquad \text{for each } y \in Y \tag{20}$$

which evidently corresponds to the max-min composition (10).

Example. Let the fuzzy conditional statement be

IF (L is "low") THEN (K is "high") = IF ("low") THEN ("high") = ("low") × ("high")

which, for "low" = 1/1 + 0.7/2 + 0.3/3 and "high" = 0.2/1 + 0.5/2 + 0.8/3 + 1/4, corresponds to

$$R = (\text{"low"}) \times (\text{"high"}) = \quad x \begin{array}{c} \\ 1 \\ 2 \\ 3 \end{array} \overset{y}{\begin{array}{cccc} 1 & 2 & 3 & 4 \\ \left[0.2 \right. & 0.5 & 0.8 & \left. 1 \right] \\ 0.2 & 0.5 & 0.7 & 0.7 \\ 0.2 & 0.3 & 0.3 & 0.3 \end{array}}$$

If now L = "medium" = 0.5/1 + 1/2 + 0.5/3, then the induced K is

K = ("medium") o R = 0.2/1 + 0.5/2 + 0.7/3 + 0.7/4

For more details on the linguistic approach, see the source papers by Zadeh [17, 18].

3.2. Some remarks on applications

Fuzzy sets have found numerous applications in a variety of fields: economics, sociology, psychology, medicine, operations research and management science, computer science, technology, etc.

In most of those applications, however, fuzzy sets have been used in some linguistic setting, i.e. using linguistic (fuzzy) values of variables and parameters, and fuzzy conditional statements (or fuzzy algorithms) to describe relations between them.

Among those works, the first were in 1974 on fuzzy controllers (Mamdani [12]) and organizational structures (Kacprzyk [7]). Basically, in the first there is a technological process whose model is unknown. Process operators have accumulated however much experience in controlling the process, and their verbal testimonies (relation between process situations and a necessary control actions) are presented as fuzzy conditional statements, i.e. a fuzzy relation being a fuzzy model of the controller. If is then used to automatically control the process. Fuzzy controllers have proved to be very efficient (see, e.g. , Mamdani [13]).

The works on organizational structures concern, first, modelling of some relations and processes in organizations (Wenstop [15]), and, second, determination of an optimal human organizational (hierarchical) structure in which both the work outcome and its dependance on some employees' features may only be verbally described (Kacprzyk [7,8,9]).

Among other works on the use of the linguistic approach, Fedrizzi, Kacprzyk and Molinari's [4] work on modeling corporate decision making processes, and Straube's linguistic forcasting models should be mentioned. Straube's works have led to the development of a powerful software package for fuzzy modeling which is presented later in this volume.

4. CONCLUDING REMARKS

We hope that this brief introduction to fuzzy sets will be useful for many readers who have not been yet exposed to fuzzy sets. Those readers should profit, by learning quickly the essence of relevant elements of fuzzy sets theory which is considerably facilitated by a constructive discussion illustrated by many examples. This should also greatly help the readers to find possible applications of fuzzy tools in their specific problems.

REFERENCES

[1] Bandler, W., and L. Kohout: Fuzzy power sets and fuzzy implication operators. Fuzzy Sets & Syst. 4 (1980), 13-30.

[2] Bellman, R. E., and L. A. Zadeh: Decision-making in a fuzzy environment. Manag. Sci. 17 (1970), 141-164.

[3] Dubois, D. and H. Prade: Fuzzy Sets: Theory and Applications. Academic Press, New York, 1980.

[4] Fedrizzi, M., J. Kacprzyk, and F. Molinari: Determination of corporate investment policies in a "soft" setting using fuzzy decision making models. Proc. 3rd IFAC/IFORS Symp. on Large Scale Syst. (Warsaw, 1983). Pergamon, Oxford, 1984.

[5] Gaines, B. R.: Foundations of fuzzy reasoning. Int. J. Man-Machine Stud. 8 (1976), 623-668.

[6] Gigch, J. P. van, and L. L. Pipino: From absolute to probable and fuzzy in decision-making. Kybernetes 9 (1980), 47-55.

[7] Kacprzyk, J.: An attempt to use fuzzy linguistic variables in the description and optimization of organizational structures (in Polish). Proc. VI Nat. Conf. on Aut. Control, Poznan, 1974.

[8] Kacprzyk, J.: A fuzzy-set-theoretic approach to the optimal assignment of work places. Proc. IFAC Conf. on Large Scale Systems Theory and Applications, Udine, 1976.

[9] Kacprzyk, J.: A fuzzy-set-theoretic approach to the optimization of organizational structures. Proc. IFAC Workshop on Control and Manag. in Integr. Ind. Complexes (Toulouse, 1977) Pergamon, Oxford, 1978.

[10] Kacprzyk, J.: Multistage Decision-Making Under Fuzziness. Verlag TÜV Rheinland, Cologne, 1983.
[11] Kaufmann, A.: Introduction à La Theorie de Sous-ensembles flous, Vol. 1. Masson, Paris, 1973. (also the English version: Introduction to the Theory of Fuzzy Subsets. Academic Press, New York, 1975).
[12] Mamdani, E. F.: Applications of fuzzy algorithms for control of a simple dynamic plant. Proc. IEE 121 (1974) 1585-1588.
[13] Mamdani, E. F.: Application of fuzzy set theory to control systems: a survey. In: Gupta, M. M., G. N. Saridis, and B. R. Gaines (eds.): Fuzzy Automata and Decision Processes. North-Holland, Amsterdam, 1977.
[14] Negoita, C. V., and D. Ralescu: Applications of Fuzzy Sets to Systems Analysis. Birkhäuser, Basel, 1975.
[15] Wenstop, F.: Deductive verbal models of organizations. Int. J. Man-Machine Studies. 8 (1979), 293-311.
[16] Zadeh, L. A.: Fuzzy sets. Inf. & Control. 8 (1965), 338-353.
[17] Zadeh, L. A.: Outline of a new approach to the analysis of complex systems and decision processes. IEEE Trans. on Syst., Man and Cybern. SMC-3 (1973), 28-44.
[18] Zadeh, L. A.: The concept of a linguistic variable and its application to approximate reasoning. Inf. Sci. Part I: 8 (1975) 199-249; Part II: 8 (1975) 301-357; Part III: 9 (1975) 43-80.

FUZZY DISPOSITIONS AS A TOOL FOR MORE
HUMAN-PERCEPTION-CONSISTENT MODELS IN SYSTEMS ANALYSIS

Janusz Kacprzyk *

Abstract. The introduction of some elements of commonsense knowledge into systems analytic models is discussed. Commonsense knowledge is represented due to Zadeh as a collection of dispositions, i.e. propositions with implicit (fuzzy) linguistic quantifiers. First, we sketch two fuzzy-logic-based calculi for manipulating the dispositions. Then, we present their use in some decision making and knowledge engineering models. The approach seems to bring the models closer to their human perception, hence should facilitate implementations.

1. INTRODUCTION

Systems analysis is basically a body of mathematical, computer based tools and techniques to support decision and policy making in complex real world problems. In most of them human elements play a crucial role. Applicability of systems analytic models may be enhanced by accounting for some relevant features of the human element. However, this may be difficult due to an inherent inconsistency between "hard" conventional models, and "soft" human perceptions, value systems, etc. Some "softening" of the models is therefore advocated (e.g., Rapaport [23] or Checkland [5]), and fuzzy sets and possibility theory can provide effective means (e.g., Kacprzyk [7]).

We present here a further "softening" of fuzzy models. Basically, we introduce some elements of commonsense knowledge in the setting of Zadeh [29], i.e. equated with a collection of dipositions (propositions with implicit fuzzy linguistic quantifiers). For instance, a disposition "winter days are cold" is in fact, e.g., "most winter days are cold". The dispositions are then manipulated using some fuzzy-logic-based calculi of linguistically quantified propositions.

First, we briefly present two alternate calculi of linguistically quantified propositions which make Zadeh's approach to commonsense knowledge operational. Then, we presents their use for "softening" multicriteria, multistage and group decision making models, as well as some knowledge-engineering-related ones (e.g., "soft" data summaries and evidence aggregation).

* Systems Research Institute, Polish Academy of Sciences,
ul. Newelska 6, 01-447 Warsaw, Poland

Our notation is standard. A fuzzy set A in X, $A \subseteq X$, is characterized by, and often equated with its membership function $\mu_A: X \to [0,1]$. For a finite $X=\{x_1,\dots,x_n\}$, A is written $A=\mu_A(x_1)/x_1+\dots+\mu_A(x_n)/x_n$. Moreover, $a \wedge b = \min(a,b)$, and $\bigwedge_{i=1}^{n} a_i = a_1 \wedge \dots \wedge a_n$.

2. FUZZY DISPOSITIONS AS A REPRESENTATION OF COMMONSENSE KNOWLEDGE

Commonsense knowledge, which plays a crucial role in our next consideration, is viewed in the setting of Zadeh [29] as a collection of dispositions, i.e. propositions with implied fuzzy quantifiers. For instance, a disposition "winter days are cold" does implicitly involve a fuzzy quantifier, say "most", i.e. it should be read as "most winter days are cold". For representing and manipulating propositions with fuzzy quantifiers, Zadeh [27, 28] has developed a fuzzy logic-based calculus. A different calculus is due to Yager [24] (a similar approach is given in Lesmo, Saitta and Torasso [19]). We will sketch below both the calculi.

A linguistically quantified proposition, examplified by "most experts are convinced", can be generally written as

$$QY's \quad are \quad F \tag{1}$$

where Q is a linguistic quantifier (e.g., most), $Y=\{y\}$ is a set of objects (e.g., experts) and F is a property (e.g., convinced).

Importance B can also be introduced yielding

$$QBY's \quad are \quad F \tag{2}$$

e.g.,"most of the important experts are conviced".

Basically, the problem is to find truth (QY's are F) knowing all truth (y_i is F), $y_i \in Y$, or in case of (2) - to find truth (QBY's are F). Fuzzy logic can be used for that purpose, and two methods will be presented below.

2.1. The algebraic (consensory) method

This classical method is due to Zadeh [27, 28]. The linguistic quantifier is here assumed to be a fuzzy set in $[0,1]$, $Q \subseteq [0,1]$. For instance, Q = "most" can be given as

$$\mu_{\text{"most"}}(x) = \begin{cases} 1 & \text{for } x \geqslant 0.8 \\ 2x - 0.6 & \text{for } 0.3 < x < 0.8 \\ 0 & \text{for } x \leqslant 0.3 \end{cases} \tag{3}$$

In the paper we will use only the proportional fuzzy quantifiers, e.g., "most", "almost all", etc. For the absolute quantifiers, e.g., "about 5", much than 10", etc. the reasoning is similar.

Property F is defined as a fuzzy set in Y, $F \subseteq Y$. If $Y=\{y_1,...,y_p\}$, then truth $(y_i$ is $F)=\mu_F(y_i)$, $i=1,...,p$.

The calculation of truth (QY's are F) is based on the nonfuzzy cardinalities, Σ Counts, of the respective fuzzy sets and proceeds in two steps:

$$r = \Sigma\, \text{Count}(F)/\Sigma\, \text{Count}(Y) = \frac{1}{p} \sum_{i=1}^{p} \mu_F(y_i) \tag{4}$$

$$\text{truth (QY's are F)} = \mu_Q(r) \tag{5}$$

To introduce importance, B = "important" is defined as a fuzzy set in Y, $B \subseteq Y$, such that $\mu_B(y_i) \in [0,1]$ is a degree of importance of y_i: the higher its value the more important y_i. We rewrite "QBY's are F" as "Q (B and F)Y's are F" which leads to the following counterparts of (4) and (5):

$$r' = \sum_{i=1}^{p} (\mu_B(y_i) \wedge \mu_F(y_i)) / \sum_{i=1}^{p} \mu_B(y_i) \tag{6}$$

$$\text{truth (QY's are F)}=\mu_Q(r') \tag{7}$$

"$\wedge$" may be replaced, e.g., by any t-norm (Kacprzyk and Yager [18]).

Example 1. Let Y = "experts" = {X,Y,Z}, F = "convinced" = 0.1/X+0.6/Y+0.8/Z, Q = "most" be given by (3), B = "important" = 0.2/X+0.5/Y+0.6/Z. Then, r=0.5, r'=0.9, and truth ("most experts are convinced")=0.4, truth ("most of the important experts are convinced")=1.

Instead of the nonfuzzy cardinalities, Σ Counts, the fuzzy cardinalities, F Counts, may also be used (Zadeh [28]). Finally, let us introduce the concept of a nondecreasing quantifier defined as

$$r_1 > r_2 \Longrightarrow \mu_Q(r_1) \geq \mu_Q(r_2) \text{ for each } r_1, r_2 \in [0,1] \tag{8}$$

which is important in our context. "Most" (3) is evidently nondecreasing. The method presented may be viewed to yield a consensory like aggregation of the pieces of evidence "y_i is F" (Kacprzyk [6], Yager [24]).

2.2. The substitution (competitive) method

This althernative method is proposed by Yager [24]. (Lesmo, Saitta and Torasso [19] is also close in sprit).

As previously, $Y=\{y_1,...,y_p\}$ and $F \subseteq Y$. We denote P_i:"y_i is F", $i=1,...,p$, and

$$\text{truth } P_i = \text{truth } (y_i \text{ is F}) = \mu_F(y_i) \tag{9}$$

We introduce the set

$$V = \{v\} = 2^{\{P_1,...,P_p\}} \tag{10}$$

where {v} is $\{P_{k1},...,P_{km}\}$ written as $\{P_{k1}$ and ... and $P_{km}\}$; {v} and $\{P_{k1}$ and ... and $P_{km}\}$ will be used interchangeably. Each v is seen to have the truth

$$\mu_T(v) = \text{truth}(P_{k1} \text{ and } \ldots \text{ and } P_{km}) = \text{truth } P_{k1} \wedge \ldots \wedge \text{truth } P_{km} =$$

$$= \bigwedge_{i=1}^{m} \text{truth } P_{ki} = \mu_F(y_{k1}) \wedge \ldots \wedge \mu_F(y_{km}) = \bigwedge_{i=1}^{m} \mu_F(y_{ki}) \tag{11}$$

The truth of "QY's are F" is now

$$\text{truth (QY's are F)} = \max_{v \in V}(\mu_Q(v) \wedge \mu_T(v)) \tag{12}$$

If $B \subseteq Y$ is importance, then

$$\text{truth (QBY's are F)} = \max_{v \in V}(\mu_Q(v) \wedge \bigwedge_{i=1}^{m} (\mu_B(y_{ki}) \rightarrow \mu_F(y_{ki}))) \tag{13}$$

where "$\rightarrow$" is an implication operator (Yager [25]). If, e.g., $a \rightarrow b = (1-a) \vee b$, then

$$\text{truth (QBY's are F)} = \max_{v \in V}(\mu_Q(v) \wedge \bigwedge_{i=1}^{m} ((1-\mu_B(y_{ki})) \vee \mu_F(y_{ki}))) \tag{14}$$

Example 2. If we take the same data as in Example 1, and Q = "most" is given as

$$\mu_{\text{"most"}}(v) = \begin{cases} 1 & \text{for } v \in \{P_1 \text{ and } P_2 \text{ and } P_3\} \\ 0.7 & \text{for } v \in \{P_1 \text{ and } P_2,\ P_1 \text{ and } P_3,\ P_2 \text{ and } P_3\} \\ 0.3 & \text{for } v \in \{P_1,\ P_2,\ P_3\} \end{cases} \tag{15}$$

then

truth ("most experts are convinced") = 0.6

truth ("most of the important experts are convinced") = 0.8

Notice that the substitution method may yield different results than the algebraic method.

A particularly type of quantifier is the monotonic quantifier defined as

$$v_1 = v_2 \text{ and } v_3 \Longrightarrow \mu_Q(v_1) \geq \max(\mu_Q(v_2), \mu_Q(v_3)) \tag{16}$$

for each $v_1, v_2, v_3 \in V$. "Most" given by (15) is evidently monotonic. The monotonic quantifiers are close in spirit to the nondecreasing ones. The substitution method may be viewed to provide a competitive like aggregation of pieces of evidence (Kacprzyk [6], Yager [24]).

The presented calculi will be now applied in a variety of models in decision making and knowledge engineering. For simplicity we will use Zadeh's calculus. The use of Yager's calculus does not change the idea but evidently complicates the corresponding mathematical expressions; for details, see the respective references.

3. SOME COMMONSENSE-KNOWLEDGE-BASED MODELS FOR DECISION MAKING AND KNOWLEDGE ENGINEERING

We will now present the use of linguistically quantified statements (dispositions) for "softening" decision making and knowledge engineering models. For lack of space we will discuss in more detail multicriteria and multistage decision making, and only briefly mention group decision making and knowledge engineering.

3.1. Multicriteria decision making

Multicriteria decision making under fuzziness may be meant as follows. $A=\{a\}=\{a_1,\dots,a_q\}$ is a set of possible alternative decisions (options) and $0=\{0_1,\dots,0_p\}$ is a set of fuzzy objectives (fuzzy constraints and/or fuzzy goals). The degree to which $a\in A$ satisfies $0_i\in 0$ is given by truth $(0_i$ is satisfied (by a)$)=\mu_{0_i}(a)$.

Traditionally, it is postulated (e.g., Bellman and Zadeh,[1]) that $a\in A$ satisfy "0_1 and ... and 0_p", i.e. all the fuzzy objectives; the degree of that satisfaction is given by the fuzzy decision

$$\mu_D(a|\text{"all"})=\text{truth}(0_1 \text{ and ... and } 0_p \text{ are satisfied}) =$$

$$=\text{truth}(\text{"all" } 0\text{'s are satisfied})=\text{truth}(0_1 \text{ is satisfied})\wedge\dots$$

$$\dots\wedge\text{truth}(0_p \text{ is satisfied})=\mu_{0_1}(a)\wedge\dots\wedge\mu_{0_p}(a) \tag{17}$$

The problem is to find an optimal decision $a^*\in A$, such that

$$a^*=\arg\max_{a\in A}\mu_D(a|\text{"all"}) \tag{18}$$

i.e. which maximizes $\mu_D(\cdot|\cdot)$ with respect to a.

The requirement to satisfy "all" the fuzzy objectives may be viewed too rigid and restrictive for practical purposes. An idea to replace "all" by some milder requirement specified by a linguistic quantifier Q, say "most", appeared in Yager [24] and Kacprzyk and Yager [17, 18]. Basically, it consists of seeking an optimal decision that best satisfies Q (e.g.,"most") fuzzy objectives.
We introduce first the fuzzy sets: S="satisfied"$\subsetneq 0$, such that $\mu_S(0_i)=$ $=\text{truth}(0_i$ is satisfied (by(a)$)=\mu_{0_i}(a)$, $i=1,\dots,p$, and B="important"$\subsetneq 0$, such that $\mu_B(0_i)\in[0,1]$ is the degree of importance of objective 0_i: from 0 denoting "unimportant at all" to 1 denoting "definitely important", through all intermediate values. $Q\subsetneq[0,1]$ is a fuzzy linguistic quantifier.

The fuzzy decision is now written as

$$\mu_D(a|Q,B)=\mu_D(a|Q,\text{"important"})=\text{truth}(QB\ 0_i\text{'s are satisfied}$$

$$\text{(by a))}=\text{truth}(Q\text{ "important" } 0_i\text{'s are satisfied}) \tag{19}$$

Employing (6) and (7), we obtain

$$\mu_D(a|Q,B) = \mu_Q(\sum_{i=1}^{p}((\mu_B(0_i) \wedge \mu_{0_i}(a))/\sum_{i=1}^{p}\mu_B(0_i) \quad (20)$$

and the problem is to find

$$a^* = \arg \max_{a \in A} \mu_D(a|Q,B) \quad (21)$$

Notice that if we do not wish to account for importance, we set $\mu_B(0_i)=1$ for each $0_i \in 0$, and (20) and (21) become, respectively:

$$\mu_D(a|Q) = \mu_Q(\frac{1}{p}\sum_{i=1}^{p}\mu_{0_i}(a)) \quad (22)$$

$$a^* = \arg_{a \in A} \mu_D(a|Q) \quad (23)$$

Evidently, it is difficult to say something about the solution of (23) for an arbitrary Q. However, for a nondecreasing Q we obtain a solution by solving

$$a^* = \arg \max_{a \in A} \mu_D(a|QB) = \arg \max_{a \in A}(\sum_{i=1}^{p}(\mu_B(0_i) \wedge \mu_{0_i}(a))/\sum_{i=1}^{p}\mu_B(0_i)) \quad (24)$$

Example 3. Let: $A=\{a_1,a_2,a_3\}$, $0_1=1/a_1+0.7/a_2+0.1/a_3$, $0_2=0.2/a_1+1/a_2+0.5/a_3$, $0_3=0.2/a_1+0.3/a_2+1/a_3$, Q="most" be given by (3), and B="important"$=0.3/0_1+$ $+0.8/0_2+1/0_3$. Then, due to (22), $\mu_D(a_1|.,.)=0$, $\mu_D(a_2|.,.)=1$, and $\mu_D(a_3|.,.)=0.9$, i.e. $a^*=a_2$.

3.2. <u>Multistage decision making (control)</u>

Multistage decision making (control) under fuzziness may be formulated as follows. At each time (control stage) t, the control $u_t \in U=\{c_1,\dots,c_m\}$ is subjected to a fuzzy constraint $\mu_{C^t}(u_t)$, and on the state attained $x_{t+1} \in X=\{s_1,\dots,s_n\}$ a fuzzy goal $\mu_{G^{t+1}}(x_{t+1})$ is imposed; the state transitions are governed $x_{t+1}=f(x_t,u_t)$; x_t, $x_{t+1} \in X$, $u_t \in U$, $t=0,1,\dots,N$; N is some termination time.

It is commonly postulated (e.g., Bellman and Zadeh [1]; or Kacprzyk [7]) that at each t, u_t satisfy the fuzzy constraint C^t and the fuzzy foal G^{t+1}, to be written as P_{t+1}: "C^t and G^{t+1} are satisfied (by u_t)". This satisfaction is evidently to the degree truth $P_{t+1}=\mu_{C^t}(u_t) \wedge \mu_{G^{t+1}}(x_{t+1})$. Traditionally, we require a sequence of controls to satisfy the fuzzy constraints and fuzzy goals at <u>all</u> the subsequent control stages, hence the fuzzy decision is

$$\mu_D(u_0,\dots,u_{N-1}|x_0,\text{"all"}) = \text{truth}(P_1 \text{ and} \dots \text{and } P_N|\text{"all"}) =$$

$$=(\bigwedge_{t=0}^{N-1}|\text{"all"})\text{truth } P_{t+1}=(\bigwedge_{t=0}^{N-1}|\text{"all"})(\mu_{C^t}(u_t) \wedge \mu_{G^{t+1}}(x_{t+1})) =$$

$$= \bigwedge_{t=0}^{N-1} (\mu_{C^t}(u_t) \wedge \mu_{G^{t+1}}(x_{t+1})) \tag{25}$$

and we seek an optimal sequence of controls $u_0^*, \ldots, u_{N-1}^*$, such that

$$u_0^*, \ldots, u_{N-1}^* = \arg \max_{u_0, \ldots, u_{N-1}} \mu_D(u_0, \ldots, u_{N-1} | x_0, \text{"all"}) \tag{26}$$

For details on some extensions of the above basic formulation, see Kacprzyk [7].

As in case of multicriteria decision making, "all" may be viewed too restrictive, and its replacement by a milder linguistic quantifier Q, say "most", has been proposed by Kacprzyk [6]. We seek therefore $u_0^*, \ldots, u_{N-1}^*$ which best satisfies the fuzzy constraints and goals at Q control stages, i.e.

$$u_0^*, \ldots u_{N-1}^* = \arg \max_{u_0, \ldots, u_{N-1}} \mu_D(u_0, \ldots, u_{N-1} | x_o, Q) =$$

$$= \arg \max_{u_0, \ldots, u_{N-1}} (\bigwedge_{t=0}^{N-1} | Q)(\mu_{C^t}(u_t) \wedge \mu_{G^{t+1}}(x_{t+1})) \tag{27}$$

For simplicity, we will assume that the importance of each control stage is the same.
Similarly as for (24), it is difficult to say something about the solution of (27) for an arbitrary Q but for a nondecreasing Q we obtain a solution by solving

$$u_0^*, \ldots, u_{N-1}^* = \arg \max_{u_0, \ldots, u_{N-1}} \frac{1}{N} \sum_{t=0}^{N-1} (\mu_{C^t}(u_t) \wedge \mu_{G^{t+1}}(x_{t+1})) \tag{28}$$

We can determine $u_0^*, \ldots, u_{N-1}^*$ – or optimal policies $a_t^* : X \to U$, $u_t^* = a_t^*(x_t)$ – by dynamic programming (Kacprzyk [6]).

Example 4. Let N=3, the fuzzy constraints and fuzzy goals be:

$$C^0 = 0.5/c_1 + 1/c_2; \; C^1 = 1/c_1 + 0.7/c_2; \; C^2 = 1/c_1 + 0.6/c_2;$$

$$G^1 = 0.1/s_1 + 0.6/s_2 + 1/s_3; \; G^2 = 0.6/s_1 + 1/s_2 + 0.5/s_3; \; G^3 = 1/s_1 + 0.8/s_2 + 0.3/s_3$$

and the state transitions be governed by

$x_{t+1} = u_t$

u_t \ x_t	s_1	s_2	s_3
c_1	s_3	s_3	s_3
c_2	s_2	s_1	s_2

Solving (28) by dynamic programming (Kacprzyk [6]), we obtain: $a_2^*(s_1)=c_2$, $a_2^*(s_2)=c_2$, $a_2^*(s_3)=c_2$; next $a_1^*(s_1)=c_2$, $a_1^*(s_2)=c_2$, $a_1^*(s_3)=c_2$; and finally $a_0^*(s_1)=c_1$ or c_2, $a_0^*(s_2)=c_1$, $a_0^*(s_3)=c_1$ or c_2.

3.3. Remarks on the use of fuzzy linguistic quantifiers in group decision making

Fuzzy linguistic quantifiers has been also successfully used in group decision making (Kacprzyk [9]) meant in our context as follows. We have a set of n options, $S= s_1,\dots,s_n$, and m individuals. Each individual k provides his preferences over S, which may be not clear-cut, hence their representation by individual fuzzy preference relations (see Bezdek et al. [2], Blin [3], Blin and Whinston [4], Kacprzyk [9, 13], Nurmi [21], Tanino [22]).

An individual fuzzy preference relation of individual k, R_k, is given by its membership function $\mu_{R_k}: S \times S \to [0,1]$. If card S is small, R_k may be conveniently represented by the $n \times n$ matrix $R_k=[r^k_{ij}]$, $ij=1,\dots,n$, $k=1,\dots,m$, whose elements $0 \leq r^k_{ij} \leq 1$ are such that the higher r^k_{ij}, the higher the preference of individual k of s_i over s_j. Moreover, it is usually assumed that the above matrix is reciprocal, i.e. $r^k_{ij}+r^k_{ji}=1$; and $r^k_{ii}=0$. Now, given $\{R_1,\dots,R_m\}$, we seek an option (or a set of options) which is "best" acceptable by the group. Basically, in our context we may proceed due to: (1) a direct approach, i.e. $\{R_1,\dots,R_m\} \to$ solution; or (2) an indirect approach, i.e. $\{R_1,\dots,R_m\} \to R \to$ solution, where R is a social fuzzy preference relation that is similarly defined as an individual fuzzy preference relation but concerns the preferences of the group (Blin and Whinston [4]).

The solution is here not clearly understood, hence varous solution concepts exist - see, e.g., Nurmi [21]. Basically they contain options that are "best" acceptable by some majority of individuals. As opposed to conventional approaches in which majority is usually a crisp threshold number of individuals, in Kacprzyk [9, 12, 13] majority is proposed to be fuzzy, given by a linguistic quantifier Q, (e.g., "most").

Using such a fuzzy majority, counterparts of some solution concepts discussed in Nurmi [21] have been introduced (Kacprzyk [9, 13]). Basically, for the direct approach these are the fuzzy Q-core, fuzzy α/Q-core and fuzzy s/Q-core which are fuzzy sets of options nondomianted, sufficiently (at most to the degree α) nondominated and strongly nondominated by Q individuals. For the indirect approach, the concepts of the fuzzy Q-consensus winner, fuzzy α/Q-consensus winner and fuzzy S/Q-consensus winner are introduced as fuzzy sets of options which are not dominated, sufficiently (at most to the degree α) dominated and strongly dominated, respectively, by other options.

In the context of group decision making linguistic quantifiers are also use to determine a degree of consensus (Kacprzyk [12]).

3.4. Remarks on the use of fuzzy linuistic quantifiers in knowledge engineering

It is easy to see that the use of linguistic quantifiers in the decision making models mentioned above virtually serves the purpose of "softening" the aggregation of partial scores (outcomes). This is also important in a multitude of problems in a broadly perceived knowledge engineering.

First, the application of linguistic quantifiers to "soft" data summaries (Yager [26]) should be mentioned. Basically, from a (nonfuzzy) data base we try in this case to find the truth of, say, "most of the items have large values". Second, two applications to evidence aggregation are proposed, one in a more general setting (Lesmo, Saitta and Torasso [20]), and one for emergency detection and monitoring (Kacprzyk and Yager [15]). In the second we basically try to find a degree of emergency if the notion of emergency is not clear-cut and is defined as a hierarchical arrangement of (possibly overlapping) collections of emergency indices and indicators.

4. CONCLUDING REMARKS

In the paper we present and advocate the use of some elements of commonsense knowledge into a variety of decision making and knowledge engineering models. This should make the models more human consistent, hence closer to reality and easier implementable. As a tool we use Zadeh's approach in which commonsense knowledge is equated with a collection of dispositions, i.e. propositions with implicit linguistic quantifiers. The approach is simple and efficient which makes the models operational.

Importance of the models may be also viewed from another perspective. The future decision support systems based upon the 5th Generation computing technology will certainly need some new human consistent and knowledge based models. And this is what the presented models may offer.

REFERENCES

[1] Bellman R.E. and L.A. Zadeh, Decision-making in a fuzzy environment, Manag. Sci. 17, (1970), 151-169.

[2] Bezdek J., B. Spillman and R. Spillman, A fuzzy relation space for group decision theory. Fuzzy Sets and Syst. 1, (1978), 255-268.

[3] Blin J.M., Fuzzy relations in group decision theory. J. Cybern. 4, (1974), 17-22.

[4] Blin J.M. and A.P. Whinston, Fuzzy sets and social choice. J. Cybern. 3, (1973), 28-33.

[5] Checkland P.B., Towards a system-based methodology of real world problem solving. J. Syst. Eng. 3, (1972).

[6] Kacprzyk J., A generalization of fuzzy multistage decision-making and control via linguistic quantifiers. Int. J. Control, 38, (1983), 1249-1270.

[7] Kacprzyk J., Multistage Decision-Making under Fuzziness: Theory and Applications, Verlag TÜV Rheinland, Cologne, 1983.

[8] Kacprzyk J., Some "commonsense" solution concepts in group decision making using fuzzy linguistic quantifiers. In [16] (forthcoming).

[9] Kacprzyk J., Collective decision making with a fuzzy majority. Proc. 5th WOGSC Congress, Paris, 1984.

[10] Kacprzyk J., (Guest Ed.), Special Issue on Fuzzy Sets and Possibility Theory in Optimization Models, Control and Cybern., 13 (1984) 4.

[11] Kacprzyk J., A generalized formulation of fuzzy multistage decision making and control under fuzziness. In Sanchez E. and M.M. Gupta (Eds.), Proc IFAC Symp. on Fuzzy Inf., Knowledge Representation and Dec. Anal., Marseille, Pergamon, Oxford, 1984.

[12] Kacprzyk J., On some fuzzy cores and "soft" consensus measures in group decision making. In Bezdek J. (Ed.), Proc. FIP-84, Hawaii, CRC Press (forthcoming).

[13] Kacprzyk J., Group decision making with a fuzzy linguistic majority, Fuzzy Sets and Syst. (forthcoming).

[14] Kacprzyk J., Towards more "human consistent" multistage decision making and control models using fuzzy sets and fuzzy logic, In Esogbue A.O. (Guest Ed.), Special Bellman Memorial Issue, Fuzzy Sets and Syst. (forthcoming).

[15] Kacprzyk J. and R.R. Yager, Emergency-oriented expert systems: a fuzzy approach. 16th Hawaii Int. Conf. on Syst. Sci., Honolulu, 1983.

[16] Kacprzyk J. and R.R. Yager (Eds.), Fuzzy Sets and Possibility Theory for Computer-Aided Management Support, Verlag TÜV Rheinland, Cologne (forthcoming).

[17] Kacprzyk J. and R.R. Yager, "Softer" optimization and control models via fuzzy linguistic quantifiers, Inf. Sci. 34 (1984), 157-178.

[18] Kacprzyk J. and R.R. Yager, Linguistic quantifiers and belief qualification in fuzzy multicriteria and multistage decision making, In [12], (1984), 155-174.

[19] Lesmo L., L. Saitta and P. Torasso, Evaluation of quantified expressions in expert systems. Proc. 5th WOGSC Congress, Paris, 1984.

[20] Lesmo L., L. Saitta and P. Torasso, Aggregating evidence in expert systems, Int. J. on Man-Machine Stidies (forthcoming).

[21] Nurmi H., Approaches to collective decision making with fuzzy preference relations, Fuzzy Sets and Systm. 5 (1981), 187-198.

[22] Tanino T., Fuzzy preference orderings in group decision making, Fuzzy Sets Syst. 12 (1984) 117-131.

[23] Rapaport A., Modern systems theory - an outlook for coping with change. In General Systms Yearbook, Soc. for Gen. Syst. XV (1970).

[24] Yager R.R., Quantifiers in the formulation of multiple objective decision functions. Inf. Sci. 31 (1983) 107-139.

[25] Yager R.R., On the implication operator in fuzzy logic, Inf. Sci. 31 (1983) 141-164.

[26] Yager R.R., A new approach to the summarization of data, Inf. Sci. 28 (1982) 69-86.

[27] Zadeh L.A., A theory of approximate reasoning, In Hayes J.E., D. Michie and L.I. Mikulich (Eds.), Machine Intelligence 9, Wiley, New York, 1979, 149-193.

[28] Zadeh L.A., A computational approach to fuzzy quantifiers in natural languages, Comp. and Maths. with Applis. 9 (1983) 149-184.

[29] Zadeh L.A., A theory of commonsense knowledge. Memo. UCB/ERL M83/27 University of California, Brekeley (1983).

FUZZY APPROACH TO MULTIPLE CRITERIA DECISION-MAKING

Reinhard Böttner[1)]

1. Introduction

Real decision situations are commonly characterized by
- multiple, usually conflicting, criteria or attributes,
- subjective uncertain conception on preferences (of the decision maker)
- uncertain, imprecise information on data of the models, describing the decision situation.

Fuzzy set theory is a suitable tool to model, analyze and solve Multiple Criteria Decision-Making (MCDM) problems loaded with subjective, imprecise information, which differs from random uncertainty (An overview over statistical approaches to decision analysis can be found in RAIFFA (1968) or KEENEY and RAIFFA (1976). The problems of MCDM can be broadly classified into two categories (HWANG and MASUD 1979):
- Design of the 'best' alternative over all criteria by considering the various interactions within the design constraints (Multiple Objective Decision-Making, MODM)
- Selection of the 'best' alternative from usually a limited number of predetermined alternatives with the help of inter- and intra-attribute comparisons (Multiple Attributive Decision-Making, MADM).

BELLMAN and ZADEH (1970) made an essential contribution to conceptionalization and application of fuzzy sets in the field of decision-making. Their definitions of fuzzy goals and fuzzy constraints as fuzzy sets in the space of alternatives and of a fuzzy decision as the intersection of fuzzy goals and fuzzy constraints are a starting point of one of the main streams of thinking on fuzzy decision-making up to the present days. Critical steps in this approach are the
- selection (or measurement) of type and shape of the describing membership functions to model uncertain information on data and preferences
- selection of the type of connective or aggregation operator for fuzzy sets, formalizing the human aggregation and compensation behavior with respect to constraints, criteria and preferences in concrete decision making situations.

Besides the approach of BELLMAN and ZADEH (1970) there exist other methods of taking into account fuzzy information on preferences of the

---

1) Central Institute of Cybernetics and Information Processes (ZKI) of the Academy of Sciences of the GDR, DDR-1080 Berlin, Kurstraße 33

decision-maker and on the data of models describing real systems. ORLOVSKI (1978, 1981) uses the concept of a fuzzy binary preference relation specified on a set of alternatives to model the decision-making situation. The relations represent the decision-makers subjective pairwise comparisons of alternatives with respect to their multiple attributes. He develops on this basis fuzzy utility functions and a theory of a general fuzzy mathematical programming problem.

Another approach is based on deterministic goal and constraint functions containing fuzzily specified parameters. A first step in this direction was done by SOYSTER (1973) with set-valued parameters in the constraints. ORLOVSKI (1981) elaborated this mathematical programming with fuzzy parameters and reduced it to a general fuzzy mathematical programming problem.
The last two approaches unfortunately lead to a large increase of the dimension of the problem, require sophisticated, time consuming question-answering procedures which sometimes exceed the cognitive and information processing-capacity of the decision maker, especially in the case of many attributes.

Therefore we will continue the BELLMAN-ZADEH approach for the MODM and MADM situations. In this way fuzzy inter- and intra-attribute preference information directly is taken into account according to the concrete decision situation by the interactive choice of the shape of membership functions and suitable aggregation operators with a variable degree of compensation between the different criteria or constraints, taking its pattern from experimental investigations of ZIMMERMANN and ZYSNO (1980, 1983).
As a result a deterministic goal- or utility function will be generated without pairwise comparisons of alternatives also for a large number of attributes or criteria by the consideration of fuzzy empirical knowledge on human aggregation behaviour in the decision making process and of theoretical demands for properties of the final decisions (e. g. Pareto optimality).
This fuzzily derived utility function serves as basis for the transformation of the fuzzy formulated MCDM-problem in an equivalent deterministic optimization problem which can be effectively solved by the powerful algorithms of mathematical programming. Our final goal is the design and development of software for a flexible, interactive decision support system to aid in the rational choice of "good" alternatives using also fuzzy expert-information .

2. Fuzzy Multiple Objective Decision-Making (FMODM)

2.1. Description of constraints and criteria by fuzzy sets

The MODM problem is characterized by a set of constraints defining the set of alternatives E_0 and a set of quantifiable objectives (criteria) scaled e. g. in such a way that a maximization is desirable:

$$E_0 = \left\{ x \in R^m \mid g_j(x) \leqq C_j \, , \quad j = 1,2,\ldots,p \right\} \tag{1}$$

$$\max \left\{ f_1(x),\, f_2(x),\, \ldots,\, f_n(x) \right\} \tag{2}$$

In the fuzzy MODM case the boundaries of the set E_0 will be made soft by introducing ranges $d_j > 0$, in which the fulfilment of the constraints decrease from the degree 1 to zero. The constraints $g_i(x) \leqq C_j+d_j$ and the objectives $f_i(x)$ are described by membership functions f_C and f_O:

$$E_0^+ = \left\{ x \in R^m \mid g_j(x) \leqq C_j+d_j \, , \quad j = 1,2,\ldots,p \right\} \tag{3}$$

$$f_{C_j}(g_j(x)) : \left\{ g_j(x) \mid x \in E_0^+ \right\} \longrightarrow [0,1] \tag{4}$$

$$f_{O_i}(f_i(x)) : \left\{ f_i(x) \mid x \in E_0^+ \right\} \longrightarrow [0,1] \tag{5}$$

The fuzzy decision $f_E(x)$ is defined as the intersection of the fuzzy constraints and objectives:

$$f_E(x) = \bigcap_{i=1}^{n} f_{O_i}(x) \cap \bigcap_{j=1}^{p} f_{C_j}(x) \tag{6}$$

As a solution $\bar{x}$ of the fuzzy MODM-problem we consider a point in the space of alternatives at which the membership function of the fuzzy decision $f_E(x)$ attains its maximum value.

$$\bar{x} = \arg \max_{x \in E_0^+} f_E(x) \tag{7}$$

The selection of this maximizing solution takes place in order to restrict the computing expense. In the case of fuzzy MADM decision situation we also consider another conception of solution.

2.2. Selection of intersection (aggregation) operators

Already BELLMAN and ZADEH (1970) pointed out the problem of interpretation of the intersection of fuzzy sets by hard and soft 'and'.

ZIMMERMANN (1976) first applied the conceptual framework of decision making in a fuzzy environment to the solution of optimization problems and suggested empirical work to determine how decision makers employ the terms 'and' and 'or' in the context of optimization. Recent results of ZIMMERMANN and ZYSKO (1980, 1983) indicate, that human decision making behavior does not correspond to the verbal and logic connectives 'and' and 'or', represented algebraically by the min- and max-operator. Instead the investigations confirmed that people often use partially compensatory connectives, that means they allow some trade-off between different goals and constraints.
We propose in accordance with ZIMMERMANN and ZYSNO (1980, 1983) to use the min-operator as a limitational case for the noncompensatory aggregation:

$$f_E^{nc}(x) = \min \left\{ f_{O_1}(x),\ f_{O_2}(x),\ \ldots,\ f_{C_p}(x) \right\} \qquad x \in E_o^+ \qquad (8)$$

For the other extreme case, the 'fully' compensatory aggregation we prefer the weighted arithmetic mean with the weighting coefficients α_i, β_j reflecting the relative importance of the constituent terms:

$$f_E^{c}(x) = \frac{1}{n+p} \left(\sum_{i=1}^{n} \alpha_i f_{O_i}(x) + \sum_{j=1}^{p} \beta_j f_{C_j}(x) \right) \qquad x \in E_o^+ \qquad (9)$$

$$\sum_{i=1}^{n} \alpha_i + \sum_{j=1}^{p} \beta_j = 1$$

This operation seems to us a more realistic model of human compensatory behavior than the often used max-operation, because it avoids problems of ambiguity, leads to manageable equivalent deterministic optimization problems (see section 2.4.), allows for a controlling of compensation degree between different goals and constraints, possesses desirable mathematical requirements of continuity, strict monotonicity, commutativity. The disagreement with the truth tables of dual logic is no serious draw back in our opinion, since the dual 'and' and 'or' connectives are no good models for real decision-making behaviour and the formal accordance with dual logic is not required in the sequel.

For the adaption to realistic decision situations characterized by compensation behaviour of the decision-maker lying between the above two extreme cases, we propose to control the compensation degree by a parameter λ in the following linear combination expression for the fuzzy decision:

$$f_E^{\lambda}(x) = (1-\lambda)\, f_E^{nc}(x) + \lambda\, f_E^{c}(x) \qquad x \in E_o^{+} \tag{10}$$

$$0 \leqq \lambda \leqq 1$$

If the compensation parameter $\lambda = 0$, than we have noncompensatory aggregation (8), for $\lambda = 1$ the other extreme case, compensatory aggregation (9), is attained.
The parameter λ is analogous to the compensationsparameter γ in the γ-model of ZIMMERMANN and ZYSNO (1980):

$$\mu_E(x) = \left(\prod_{i=1}^{m} \mu_i^{\delta_i}(x)\right)^{1-\gamma} \cdot \left(1 - \prod_{i=1}^{m} (1-\mu_i(x))^{\delta_i}\right)^{\gamma} \tag{11}$$

$$0 \leqq \mu_i(x) \leqq 1\,, \quad 0 \leqq \gamma \leqq 1$$

m - number of sets to be connected

δ_i - weights $\qquad \sum_{i=1}^{m} \delta_i = m$

In this γ-compensation operator the aggregation is expressed as the product of the intersection and union of fuzzy sets, algebraically represented by the product and algebraic sum, respectively.
For the MODM-case we prefer the λ-model (10) because it leads to equivalent deterministic optimization problems which can be better solved and it provides almost equal good approximation of empirical values, e. g. for the data of ZIMMERMANN and ZYSNO (1980).
In the MADM-case, where a selection has to take place between only a finite number of alternatives, we will also use the γ-operator as a model of human aggregation behaviour, because its complicated algebraic structure can be mastered in this case.
For both aggregation operators the numerical values of the 'grades of compensation' λ or γ have to be determined experimentally, e. g. by data-fitting methods.

2.3. Selection of the type and shape of the describing membership function

Deciding the properties of the membership function, which describe the constraints and criteria of the MODM-problem, we include in the model-building process fuzzy information on model parameters and the preference structure of the decision-maker.
On the other hand we also determine by this selection properties of the

solution of the fuzzy MODM-problem.
The constraints (3,4) are often described due to ZIMMERMANN (1976, 1978) by linear membership functions $a_j = 1$ of the following type:

$$f_{C_j}(g_j(x)) = \begin{cases} 1 & \text{for } g_j(x) \leqq C_j \\ \left(\frac{C_j+d_j-g_j(x)}{d_j}\right)^{a_j} & \text{for } C_j \leqq g_j(x) \leqq C_j+d_j \\ 0 & C_j+d_j \leqq g_j(x) \end{cases} \tag{12}$$

The objective funtions may be described in the same way as fuzzy constraints with the aspiration value $\bar{f}_i$, and the range d_i in which the fulfilment of the goal decreases from 1 to zero:

$$f_{O_i}(f_i(x)) = \begin{cases} 1 & \text{for } f_i(x) \geqq \bar{f}_i \\ \left(\frac{f_i(x)+d_i-\bar{f}_i}{d_i}\right)^{a_i} & \text{for } \bar{f}_i \geqq f_i(x) \geqq \bar{f}_i-d_i \\ 0 & \bar{f}_i-d_i \geqq f_i(x) \end{cases} \tag{13}$$

with $a_i = 1$ in the linear case and it holds for the aspiration value

$$\bar{f}_i \geqq \min_{x \in E_0^+} f_i(x) \tag{14}$$

Here we have full symmetry between goals and constraints and an objective is fulfilled, if the aspiration level $\bar{f}_i$ is reached. In some application cases, e. g. nuclear risk analysis, there may be no admissible violation of the constraints and aspiration levels of objectives and one has to choose $d_i = 0$, that means to use crisp constraints with a threshold membership function.
As in the case of goal programming (KORNBLUTH 1973) the Pareto-optimality of the maximizing solution $\bar{x}$ (7) cannot be guaranted by describing the objectives as fuzzy constraints.
If we select the following type of strictly monotone objective describing membership function f_{O_i}, this drawback can be overcome.

$$\bar{f}_{O_i}(f_i(x)) : \left[\bar{f}_i \, , f_i^{max}\right] \longrightarrow [0,1] \tag{15}$$

$\bar{f}_{O_i}$ strictly monotone increasing

$$f_i^{max} = \max_{x \in E_0^+} f_i(x)$$
$$i = 1,2,\dots,n$$
$$\bar{f}_i \geqq f_i^{min} = \min_{x \in E_0^+} f_i(x)$$

$$f_{0_i}(f_i(x)) = \begin{cases} \bar{f}_{0_i}(f_i(x)) & \text{for } \bar{f}_i \leqq f_i(x) \leqq f_i^{max} \\ 0 & \text{else} \end{cases} \tag{16}$$

$\bar{f}_i$ is the minimal achievment of the i-th objective the decision-maker wants to attain.
Usually the membership function is normalized as follows:

$$f_{0_i}(\bar{f}_i) = 0 \, , \quad f_{0_i}(f_i^{max}) = 1 \tag{17}$$

FENG (1983) proved that the maximal solution $\bar{\bar{x}}$ (7) is weak efficient if the functions $f_{0_i}(x)$ ($i = 1,2,\dots,n$) are strictly monotone increasing and the maximal solution is efficient if the functions $f_i(x)$ ($i = 1,2,\dots,n$) are strictly concave and the constraint functions $g_j(x)$ ($j = 1,2,\dots,p$) are concave.
Fuzzy preference information of the decision-maker on the relative importance of the single objectives can be integrated in the model building by a suitable choice of the shape of the describing membership function.
In case of maximization of objective i the decision-maker can express a high importance of the i-th objective (that means small trade-off between the i-th and the other objectives) by the following convex membership function:

$$f_{0_i}(f_i(x)) = \begin{cases} \left(\dfrac{f_i(x)-\bar{f}_i}{f_i^{max}-\bar{f}_i}\right)^{a_i} & \text{for } f_i(x) \geqq \bar{f}_i \\ 0 & \text{for } f_i(x) \leqq \bar{f}_i \end{cases} \tag{18}$$

$$a_i \geqq 2$$

A small decrease of $f_i(x)$ gives rise to a drastic decrease of the membership function. If the decision-maker assigns only weak importance to the i-th objective and he is willing to accept trade-off with other objectives, then this situation can be modelled by a concave membership function. Now we have to choose in (18)

$$a_i = \frac{1}{n_i} \qquad n_i \geqq 2 \tag{19}$$

In this case, a strong decrease of objective values $f_i(x)$ results in a weak decrease of the value of the membership function (18). When the decision-maker is not able to assign a strong or weak importance to the i-th objective, this indifferent behaviour can be described by a linear membership function. This means in (18) $a_i = 1$.
The determination of the minimal desired objective values $\overline{f}_i$ (aspiration levels) and the shape of the membership function has to be realized in an interactive manner with the decision-maker, modelling thereby his fuzzy intention about trade-offs between criteria.

2.4. Equivalent deterministic optimization problems

In case of noncompensatory aggregation with the min-operator (8) and the concept of maximization of the membership value of the fuzzy decision, the optimal solution (7) can be found by the following mathematical programming problem (ZIMMERMANN 1976)

$$\begin{aligned} & \max x_{m+1} \\ & x_{m+1} \leqq f_{O_i}(x) \qquad i = 1,2,\ldots,n \\ & x_{m+1} \leqq f_{C_j}(x) \qquad j = 1,2,\ldots,p \\ & x_{m+1} \geqq 0 \end{aligned} \tag{20}$$

x_{m+1} is a global variable, which indicates how good the objectives and constraints are simultaneously fulfilled. It is also possible to introduce a second auxiliary variable x_{m+2} and to weight the importance of the constraint- and objective-fulfilment by the parameter w (BÖTTNER 1983).

$$\begin{aligned} & \max (w \cdot x_{m+1} + (1-w) \cdot x_{m+2}) \\ & x_{m+1} \leqq f_{O_i}(x) \qquad i = 1,2,\ldots,n \\ & x_{m+2} \leqq f_{C_j}(x) \qquad j = 1,2,\ldots,p \\ & x_{m+1} \geqq 0 \\ & x_{m+2} \geqq 0 \end{aligned} \tag{21}$$

In this way the decision-maker can express his preferences regarding the degree of goal accomplishment at the expense of degree of constraint fulfilment .

If a problem contains crisp constraints or objective functions with crisp aspiration values, one can add them in their original form directly to the equivalent deterministic problems (20,21). This is valid also for the subsequent cases.

If linear membership functions ($a_i = 1$, $a_j = 1$) $f_{O_i}(x)$ and $f_{C_j}(x)$ (18,12) are utilized the equivalent deterministic optimization problem (20) has a structure as follows:

$$
\begin{aligned}
&\max x_{m+1} \\
&x_{m+1} (f_i^{\max} - \bar{f}_i) \leqq \bar{f}_i && i = 1,2,\ldots,n \\
&x_{m+1} \cdot d_j + g_j(x) \leqq C_j + d_j && j = 1,2,\ldots,p \\
&x_{m+1} \geqq 0
\end{aligned}
\tag{22}
$$

This only means an addition of a linear variable x_{m+1} and of n constraints with a simple objective function to the original constraints. Taking into consideration weak nonlinear membership functions leads to weak nonlinear optimization problems, which can be solved effectively with the existing mathematical programming codes. For example a concave ($a_i = \frac{1}{2}$) membership function $f_{O_i}(x)$ of type (18) causes constraint equations of the following kind:

$$x_{m+1}^2 (f_i^{\max} - \bar{f}_i) - f_i(x) \leqq \bar{f}_i \tag{23}$$

In a decision making situation, in which a compensatory aggregation according to the weighted arithmetic mean (9) is adequate, the following equivalent deterministic optimization problem has to be solved:

$$\max_{x \in E_0^+} \frac{1}{n+p} \left(\sum_{i=1}^{n} \alpha_i f_{O_i}(x) + \sum_{j=1}^{p} \beta_j f_{C_j}(x) \right) \tag{24}$$

The necessary solution-effort is relatively small and depends on the selection of describing membership functions. Linear $f_i(x)$, $g_j(x)$, $f_{Oi}(x)$ and $f_{C_j}(x)$ in equations (2,3,12,18) lead to linear equivalent programming problems. Also for concave or convex membership functions ($a_i = \frac{1}{2}$, $a_j = \frac{1}{2}$ respectively $a_i = 2$, $a_j = 2$ in (12,18)) the resulting equivalent nonlinear programming problems remain in the scope of modern mathematical programming packages.

This is valid, too, for the equivalent programming problems in decision situations, which demand a controlled compensation according to the λ - model (10):

$$\max_{x \in E_o^+} \left[(1-\lambda)\, x_{m+1} + \frac{\lambda}{n+p} \left(\sum_{i=1}^{n} \alpha_i f_{O_i}(x) + \sum \beta_j f_{C_j}(x) \right)\right]$$

$$\begin{aligned} & x_{m+1} \leqq f_{O_i}(x) && i = 1,2,\ldots,n \\ & x_{m+1} \leqq f_{C_j}(x) && j = 1,2,\ldots,p \\ & x_{m+1} \geqq 0\,, \quad 1 \geqq \lambda \geqq 0 \end{aligned} \qquad (25)$$

3. Fuzzy Multiple Attributive Decision-Making (FMADM)

In the MADM decision situation one has to select the 'best' or some superior alternatives from a limited number of predetermined alternatives. The inclusion of fuzzy information in the model building and decision-making process proceeds along the same lines as in the MODM case. However the set of alternatives are not given implicitly by a set of constraints (3). Instead the set of alternatives resulting from simulation experiments, multiple objective decision procedures or other design processes can be concisely described in the following decision matrix:

	f_1	$\cdots$	f_i	$\cdots$	f_n
A_1	f_{11}	$\cdots$	f_{1i}	$\cdots$	f_{1n}
$\vdots$	$\vdots$		$\vdots$		$\vdots$
A_j	f_{j1}	$\cdots$	f_{ji}	$\cdots$	f_{jn}
$\vdots$	$\vdots$		$\vdots$		$\vdots$
A_m	f_{m1}	$\cdots$	f_{mi}	$\cdots$	f_{mn}

(26)

$\{A_j\}$ $j = 1,2,\ldots,m$ set of alternatives

$\{f_i\}$ $i = 1,2,\ldots,n$ set of criteria/attributes

Each criterion is described by a membership function of the type (15,16, 17,18,19), taking into account in this manner fuzzy preference information interactively with the decision maker.

The membership functions $f_{O_i}(f_{ji})$ are now defined only at the discrete points f_{ji} ($j = 1,2,\ldots,m$), the realizations of the m alternatives regarding the i-th criterion or attribute.

Having performed that transformations we get from the decision matrix the corresponding matrix of membership values:

$$\begin{array}{c|ccccc} & f_1 & \dots & f_i & \dots & f_n \\ \hline A_1 & f_{0_{11}} & \dots & f_{0_{1i}} & \dots & f_{0_{1n}} \\ \vdots & \vdots & & \vdots & & \vdots \\ A_j & f_{0_{j1}} & \dots & f_{0_{ji}} & \dots & f_{0_{jn}} \\ \vdots & \vdots & & \vdots & & \vdots \\ A_m & f_{0_{m1}} & \dots & f_{0_{mi}} & \dots & f_{0_{mn}} \end{array} \qquad (27)$$

For the aggregation of the different criteria we use the same concept and the corresponding operators (8,9,10,11) as in the MODM situation in the case of noncompensatory, compensatory and controlled compensatory decision-making behaviour.

To make the final decision, that means to seek the alternative with the highest degree of membership for the fuzzy decision set we do not need here sophisticated and expensive mathematical programming algorithms. Simple maximum-seeking and sorting procedures are sufficient. Thus we can even use in the FMADM-situation the algebraically complicated γ -operator (11) of ZIMMERMANN, ZYSNO (1980) for modelling human decision making behaviour.

With regard to the solution concept we are not restricted here to the maximizing solution (7). In the discrete case we can e. g. easily provide the decision-maker with level sets of the fuzzy decision (6). That means to present to him for final selection all alternatives with a membership value for the fuzzy decision set exceeding a fixed value α. For the fuzzy Multiattributive Decision-Making situation we have developed an interactive computer programm, including also compensatory aggregation according to the γ -operator of ZIMMERMANN and ZYSNO (1980). Software for decision support in the FMODM case is under development.

References

BELLMAN, R.E. and L.A. ZADEH (1970). Decision-making in a fuzzy environment. Management Science Vol. 17, No. 4 pp. B141-164

BÖTTNER, R. (1983). Fuzzy model-building and optimization and application to the analysis of complex fuel conversion (in German). ZKI-Informationen 2/83, S. 64-93

FENG, Y. (1983). A Method using fuzzy mathematics to solve the vectormaximum problem. Fuzzy Sets and Systems 9 (1983), 131-136

HWANG, C.L. and A.S. MASUD (1979). Multiple Objective Decision Making. Springer-Verlag Berlin, Heidelberg, New York 1979

KEENEY, R.L. and H. RAIFFA (1976). Decisions with multiple objectives: preferences and value trade-offs. Wiley, New York 1976

KORNBLUTH, J.S. (1973). A survey of goal programming. Omega, Vol. 1, pp. 193-205

ORLOVSKI, S.A. (1978). Decision making with a Fuzzy Preference Relation. Fuzzy Sets and Systems, Vol. 1, No. 3, pp. 155-167

ORLOVSKI, S.A. (1980). On Formalization of a General Fuzzy Mathematical Programming Problem. Fuzzy Sets and Systems, Vol. 1, No. 3, pp. 311-321

ORLOVSKI, S.A. (1981). Problems of Decision-Making with Fuzzy Information. "Nauka Publ." Moscow 1981 (in Russian)

RAIFFA, H. (1968). Decision Analysis: Introductory Lectures on Choices under Uncertainty. New York. Addison Wesley 1968

SOYSTER, A.L. (1973). Convex programming with set inclusive constraints and application to inexact linear programming. Operat. Res. 21 (1973) p. 1154-1157

ZIMMERMANN, H.J. (1976). Description and optimization of fuzzy systems. Int. J. General Systems, 2 (1976) pp. 209-215

ZIMMERMANN, H.J. (1978). Fuzzy programming with several objective functions. Fuzzy Sets and Systems, 1 (1978) pp. 46-55

ZIMMERMANN, H.J. and P. ZYSNO (1980). Latent connectives in human decision making. Fuzzy Sets and Systems, 4 (1980) pp. 37-51

ZIMMERMANN, H.J. and P. ZYSNO (1983). Decisions and evaluations by hierarchical aggregation of information. Fuzzy Sets and Systems, 10 (1983) pp. 243-260

A FLEXIBLE MATHEMATICAL PROGRAMMING SYSTEM

Brigitte Werners*

1. Introduction

One important field of application of fuzzy set theory is decision making and decision support. Vage informations and unprecise requirements are typical for complex realistic problems. Even if a problem is rather well structured, a crisp mathematical programming model is frequently too unflexible with its one goal and its strict discrimination between admissible and inadmissible solutions. Here a fuzzy approach representing informations and requirements by fuzzy sets respectively by their membership functions often is much better applicable to real situations. In a fuzzy mathematical programming model multiple objectives or flexible regions of feasible solutions can be dealt with. One main advantage of these approaches is that under certain conditions equivalent crisp models can be formulated which are solvable by efficient standard software. Generally it is assumed that the decision maker is able to indicate membership functions of his goals and constraints a priori. In the following explanation this assumption is dropped. Membership functions of goals are proposed by the system and the decision maker has the opportunity to modify them and the membership functions of constraints interactively.

According to Bellman, Zadeh [1] a fuzzy decision is the fuzzy set resulting from intersection of fuzzy goals and fuzzy constraints. An optimal decision is an element with maximal degree of membership to this set. The mathematically easiest way to model intersection is the minimum operator. Though the minimum is not always the empirically best model [6] here we will restrict on it because when this operator is used very efficient algorithms are applicable as will be shown later on.

In case all membership functions are known goals and constraints are handled fully symmetrically. So we consider the main difference between an objective - here synonymously used to goal - and a constraint depending on the determination of the membership function. In classical programming a constraint dividing all possible solutions in those which are feasible and those which are infeasible is determined by the decision maker in advance. An objective is a 'requirement which is to be followed to the greatest extend possible' [4] with regard to the constraints. Now a generalized definition can be formulated for fuzzy programming models which as is well known contain crisp mathematical programming models as special cases. A constraint is defined by its membership function which can be determined

* Brigitte Werners, Lehrstuhl für Unternehmensforschung, RWTH Aachen, Templergraben 64, D-5100 Aachen

a priori. For an objective the requirement 'as much/little as possible' depends on the constraints and no membership function can be given by the decision maker. Therefore the system proposes membership functions for objectives with regard to fuzzy constraints. The evaluation is shown in the second chapter considering one objective and is then generalized for multiple objectives.

The treatment of multiple objectives under fuzzy constraints further requires a generalization of the definition 'efficient solution', which is done in 2.2. The system determines fuzzy-efficient extreme solutions and a fuzzy-efficient compromise solution. They are judged by the decision maker and he decides wether one of them is subjectively the best one or wether modifications are necessary. In the latter case the decision maker changes membership functions assisted by the system. The following rough flow chart sketches how the decision support system works.

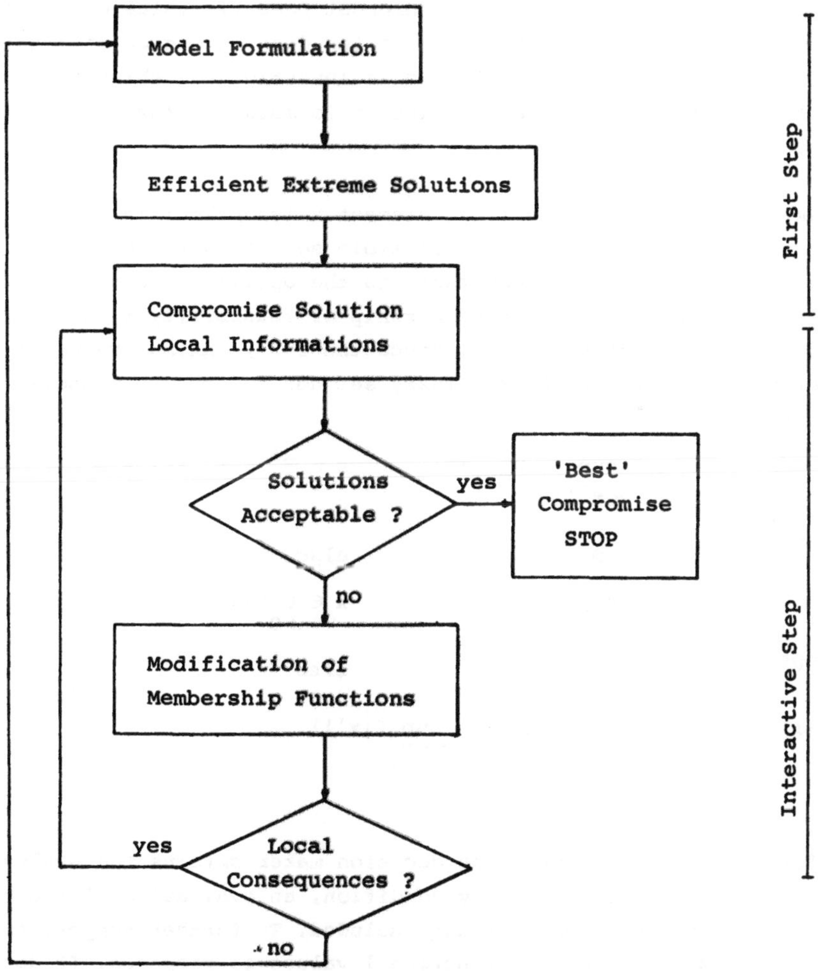

Fig. 1.1: Flow chart decision support system

2. Membership Functions of Goals under Fuzzy Constraints

2.1 One Objective Function

While formulating fuzzy programming models the decision maker is often expected to be able to specify degrees of membership for different values of different objective functions. The consequence is the perfect symmetry of goals and constraints. Yet the decision maker does not always have a clear idea of his goals but takes into account what he can achieve at the most under the given circumstances. When he uses the formulation of a classical programming model, he considers a solution as completely satisfying his goal, if this solution maximizes the objective function subject to the constraints. He has not yet a precise conception, which value the objective function will reach. The additional information obtained from the optimal solution thus causes the decision maker to transform his previous maximization idea into a satisfying conception. Provided some of the constraints of a model are fuzzy, the highest possible value of an objective function is different corresponding to the desired degree of feasibility of the solution. In the following it is investigated, which information is usefull to support the decision maker to formulate a fuzzy goal and its membership function.

Orlovsky [2] suggests: only if an element x is optimal in R_α, being the α-level-set of the fuzzy region of feasible solutions for at least one α, it gets a positive degree of membership to the optimization criteria. And that is the highest degree of membership of feasibility α, for which optimality of x can be stated in R_α. Hence the fuzzy maximal value of f(x) subject to the fuzzy set R is a fuzzy set defined by the membership function $\mu_f(r)$:

$$\mu_f(r) = \begin{cases} \sup\limits_{x \in f^{-1}(r)} \mu_{opt}(x) & r \in \mathbb{R}^1 \wedge f^{-1}(r) \neq \emptyset \\ 0 & \text{else} \end{cases} \tag{2.1.1}$$

with

$$\mu_{opt}(x) = \begin{cases} \sup\limits_{N(\alpha) \ni x} & x \in \bigcup\limits_{\alpha > 0} N(\alpha) \\ 0 & \text{else} \end{cases}$$

$$N(\alpha) = \{x \mid x \in R_\alpha \wedge f(x) = \sup_{x' \in R_\alpha} f(x')\}$$

Thus for every real number r the decision maker obtains the evaluation of the membership to the optimality condition, an evaluation of the hight of the value of the function f is not included. To further support the decision maker a fuzzy set on the functional values is proposed. It does not seem reasonable if the decision maker measures his evaluation at the lowest possible value of f subject to the constraints as is done in a maximizing set. He should rather take the highest value, which is attainable, if the constraints are satisfied with degree of membership 1, as a lower bound for positive membership to the goal. Therefore the following membership function is suggested for the values of the objective function.

Definition 2.1.1

Let $f: X \to \mathbb{R}$, and R be a fuzzy constraint set. Then the membership function of the objective function f subject to the fuzzy region of feasible solutions is defined by

$$\mu_Z(x) = \begin{cases} 0 & f(x) \le \sup_{R_1} f \\ \dfrac{f(x) - \sup_{R_1} f}{\sup_{S(R)} f - \sup_{R_1} f} & \sup_{R_1} f < f(x) < \sup_{S(R)} f \\ 1 & \sup_{S(R)} f \le f(x) \end{cases}$$

Here R_1 is the α-level-cut of R with $\alpha = 1$.

Then we get the following membership function in the objective function space:

$$\mu_Z(r) := \begin{cases} \sup_{x \in f^{-1}(r)} \mu_Z(x) & r \in \mathbb{R} \wedge f^{-1}(r) \neq \emptyset \\ 0 & \text{else} \end{cases} \qquad (2.1.2)$$

$$\mu_D(r) = \mu_f(r) \wedge \mu_Z(r),$$

the optimal value of the objective function can be obtained from

$$\sup_{r \in R} \mu_D(r).$$

The fuzzy decision in X thus is determined with $x \in f^{-1}(r)$ by

$$\mu_D(x) := \mu_D(f(x)).$$

Hence the evaluation of an objective value depends on what can be achieved subject to the given constraints, corresponding to the evaluation in classical programming models.

For a linear fuzzy model (see [3]) the membership functions $\mu_Z(r)$ and $\mu_f(r)$ are represented. By applying the minimum operator the optimal objective value is $r^o = 11{,}73$ and the degree of membership to the fuzzy constraints is 0,53, the optimal solution $x^o \in f^{-1}(r^o)$ can easily be determined.

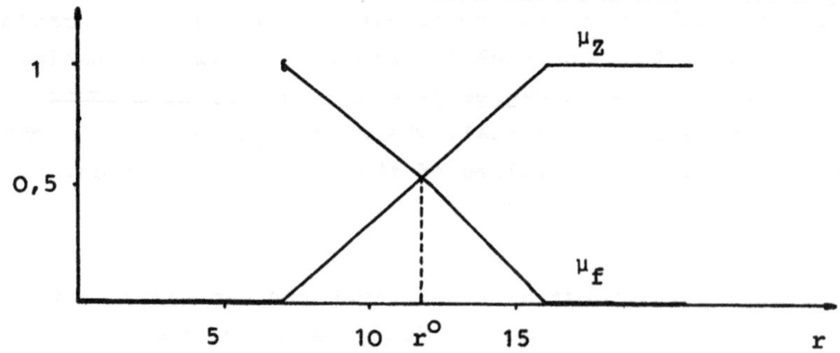

Fig. 2.1.1: Membership function of the objective function and the optimization requirement

If the decision maker accepts the formulation of the membership function of the objective function dependent on the fuzzy region of feasible solutions goal and constraints will be treated equally afterwards. In the following we shall investigate to what extend this approach can be transferred to models with multiple objective functions.

2.2. Multiple Objective Functions

Considering models with multiple goals and crisp constraints fuzzy models can be formulated to determine compromise solutions [5]. For this purpose the goal requirements are modelled using membership functions. Under additional assumptions linear substitutional models can be created. It is to be understood as a model without additional information from the decision maker and is most likely comparable with goal programming models [3].

In case the constraint set is described not only by crisp but additionally or exclusively by fuzzy constraints, the varying degrees of feasibility should be taken into account by consideration of efficient solutions, for additional dependencies between the individual goal values and the degree of membership to the region of feasible solutions can arise.

$$\begin{aligned} &\text{"max" } Cx \\ &\text{s.th. } Ax \lesssim b \\ &\qquad\quad x \geq 0 \end{aligned} \qquad (2.2.1)$$

Thus a fuzzy-efficiency is defined, which compises the classical efficiency as special case, but considers varying degrees of feasibility.

Definition 2.2.1

Let $X \subset \mathbb{R}^{m'}$, and k the number of functions $f_l: X \to \mathbb{R}^1$, $l = 1,\ldots,k$, which have to be maximized. Let m be the number of membership functions to the constraints $\mu_i: X \to [0,1]$, $i = 1,\ldots,m$. An element x is called fuzzy-efficient, if no $x' \in X$ exists with

$$\begin{aligned}
& f_l(x') \geq f_l(x) && \forall l = 1,\ldots,k \\
\wedge\ & \mu_i(x') \geq \mu_i(x) && \forall i = 1,\ldots,m \\
\wedge\ & (\ f_{l_o}(x') > f_{l_o}(x) && \text{for at least one } l_o \in \{1,\ldots,k\} \\
& \vee\ \mu_{i_o}(x') > \mu_{i_o}(x) && \text{for at least one } i_o \in \{1,\ldots,m\}).
\end{aligned}$$

This definition takes into account that for a fuzzy-efficient solution an improvement concearning an objective function can only be reached either at the expense of the membership to the constraints or at the expense of an additional objective function. For example it can happen, that an element x, for which all objective values $f_l(x)$, $l=1,\ldots,k$ are less than those of an element x', is yet fuzzy-efficient, provided the membership of x to the constraints is higher.

Fig. 2.2.1 shows a fuzzy region of feasible solutions, in which the section with membership values between 0 and 1 is hatched. The crisp objective functions are $f_1(x) = x_1$, $f_2(x) = x_2$.

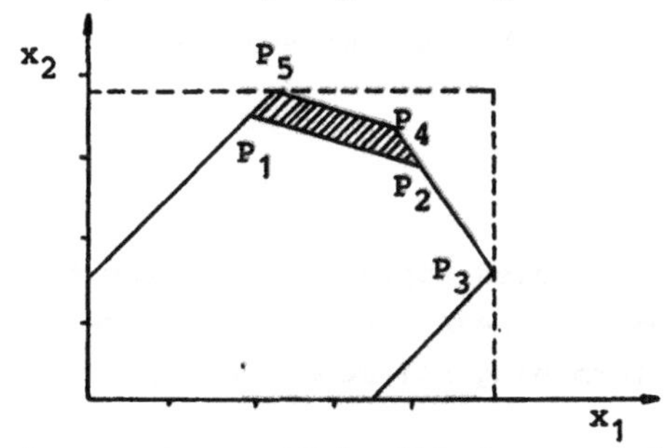

Fig. 2.2.1: Vector maximum model with fuzzy region of feasible solutions

All points on the line $\overline{P_2P_3}$ and all convex combinations of P_1, P_2, P_4 and P_5, that is the hatched section including the marginal lines, are fuzzy-efficient.

2.3 Extreme Solutions

In multiple objective problem solving methods the decision maker is frequently supported by a table of results of extreme solutions. Thus he gets a first impression of the interdependences of objective functions and constraints. For fuzzy mathematical models an extension of this table is necessary due to the fuzzy constraints. It should simplify the decision maker's acces to the process of problem solving.

For each objective function two individual optima are calculated, first under consideration of the constraint set C_R^o with positive membership, the margin being included, second subject to the constraint set C_R^1 with complete membership. C_R^1 is determined by the most restricting conditions to the constraints, C_R^o is determined by the least restricting ones. In each case a fuzzy-efficient solution can be obtained, if the solution space is not empty [3]. The decision maker receives the following table:

	c_1	c_2	. . .	c_k
x^{o1}	$\sum_j c_{1j}x_j^{o1}$	$\sum_j c_{2j}x_j^{o1}$	. . .	$\sum_j c_{kj}x_j^{o1}$
⋮				
x^{ok}	$\sum_j c_{1j}x_j^{ok}$	$\sum_j c_{2j}x_j^{ok}$	. . .	$\sum_j c_{kj}x_j^{ok}$
x^{11}	$\sum_j c_{1j}x_j^{11}$	$\sum_j c_{2j}x_j^{11}$	. . .	$\sum_j c_{kj}x_j^{11}$
⋮				
x^{1k}	$\sum_j c_{1j}x_j^{1k}$	$\sum_j c_{2j}x_j^{1k}$	. . .	$\sum_j c_{kj}x_j^{1k}$

Fig. 2.3.1: Table of results of the extreme solutions

3. Interactive Step

3.1. Compromise Solution

A compromise solution is calculated applying the fuzzy concept depicted in chap. 2. We have the following linear substitution model to (2.2.1) which can be formulated after determining all membership functions:

$$\begin{aligned} \max \quad & \alpha \\ \text{s.th.} \quad & (\bar{c} - \underline{c})\alpha - Cx \le -\underline{c} \\ & (\bar{b} - \underline{b})\alpha + Ax \le \bar{b} \\ & \alpha, x \ge 0 \end{aligned} \qquad (3.1.1)$$

An optimal solution (α^o, x^o) of (3.1.1) consists of a compromise solution x^o of (2.2.1) and the degree of membership of this solution to the fuzzy decision under the chosen membership functions of goals and constraints when the minimum operator is applied.

In addition to the fuzzy-efficient compromise solution x^o the decision maker is informed about the value of each objective function and each constraint and also the membership value realized for each membership function. The objective and constraint values are compared with the ideal and the pessimistic values, respectively the bounds of the flexible intervals of the constraints.

	goals			constraints		
		$\underset{\sim}{\max}$		$\underset{\sim}{\leq}$	$\underset{\sim}{\geq}$	$\underset{\sim}{=}$
	$\bar{c}_1$	...	$\bar{c}_k$	$\bar{b}_1^1$ ...	$\underline{b}_1^2$ ...	b_1^3 ...
x^o	$\sum_j c_{1j}x_j^o$	...	$\sum_j c_{kj}x_j^o$	$\sum_j a_{1j}^1 x_j^o$ ...	$\sum_j a_{1j}^2 x_j^o$ ...	$\sum_j a_{1j}^3 x_j^o$ ...
	$\underline{c}_1$	...	$\underline{c}_k$	$\underline{b}_1^1$ ...	$\bar{b}_1^2$ ...	$\underline{b}_1^3$, $\bar{b}_1^3$ ...
α^o	α_1	...	α_k	α_{k+1} ...	α_{k+m_1+1}	...

Fig. 3.1.1: Table of results of the compromise solution x^o

The membership value α^o of x^o in the fuzzy set decision is thus the minimum of all achieved membership values. Informations about further possible variations of the objective function might be obtained by calculating $\Delta\alpha(\Delta_1)$ [3]. But since no changes of the basis matrix are taken into account, this information is only of local interest.

If the decision maker accepts one of the proposed fuzzy-efficient solutions, i.e. either one of the extreme solutions or the compromise solution, as his best compromise, the decision making process is successfully concluded, otherwise the interactive modification starts.

3.2. Interactive Modifications

In the interactive step the decision maker has the opportunity to modify the membership functions of the goals or constraints. The substitutional model is changed accordingly and another compromise solution is proposed to the decision maker. This solution then is an optimal fuzzy decision considering the modified membership functions which are again aggregated by the minimum operator with equal weight. However, the change of the membership functions implicitly causes a modification of the weights. The membership functions, which are still assumed to be piecewise linear, are modified by indicating new values $\underline{c}_l$ or $\bar{c}_l$, respectively $\underline{b}_i$ or $\bar{b}_i$.

Example 3.2.1

Modifying one membership function and consequences for the optimal fuzzy decision are considered:

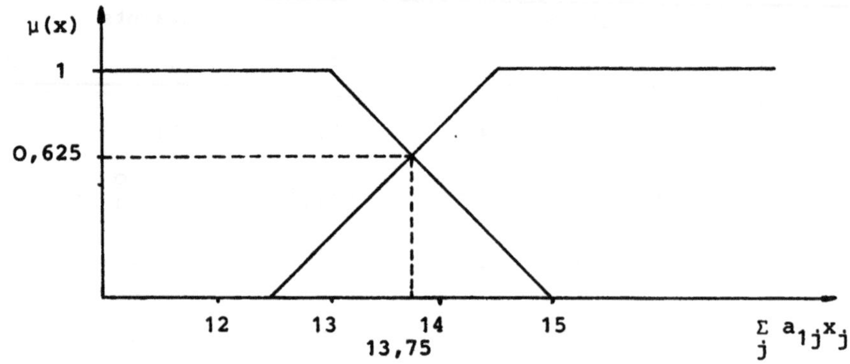

Fig. 3.2.1: Membership function with $\underline{b}_1$=12,5, $\bar{b}_1$=14,5

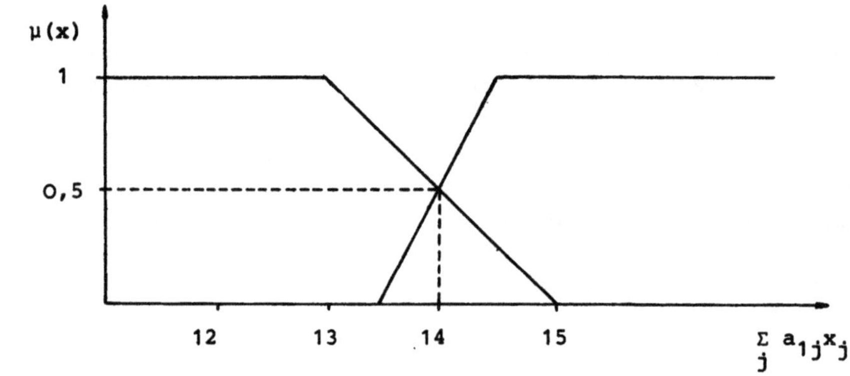

Fig. 3.2.2: Modification of $\underline{b}_1$ into $\underline{b}_1'$ = 13,5

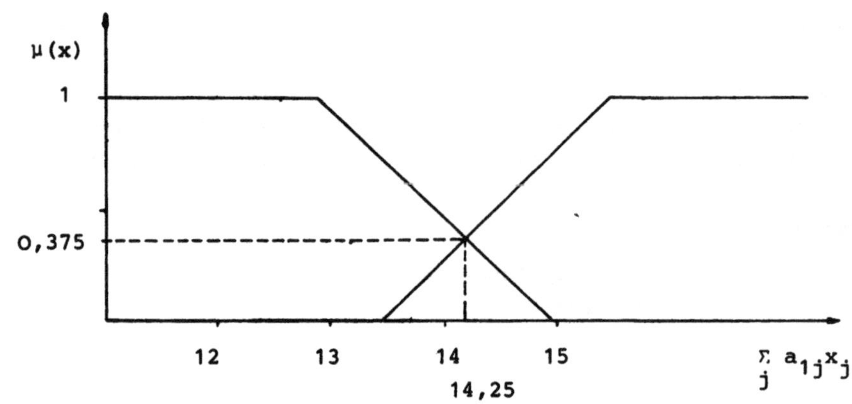

Fig. 3.2.3: Modification into $\underline{b}_1'$ = 13,5, $\bar{b}_1'$ = 15,5

In principle the modification of all these values is possible. Yet in the interactive step, when a new compromise alternative is generated, only the following varieties should be accepted without change of the model:

- goal functions: increase of $\underline{c}$
- $\underset{\sim}{\leq}$-constraints: decrease of $\bar{b}$
- $\underset{\sim}{\geq}$-constraints: increase of $\underline{b}$
- $\underset{\sim}{=}$-constraints: increase of $\underline{b}$ or decrease of $\bar{b}$.

The consequence of a decrease of $\bar{b}_i$ in $\underset{\sim}{\leq}$-constraints, for example, is that compared to the previous model, the support of the fuzzy feasible set is restricted in such a way, that a part of the former flexible intervall is not accepted any more and the flexibility is reduced.

All further modifications of the parameters $\underline{c}$, $\bar{c}$, $\bar{b}$, $\underline{b}$ should also be possible in this interactive process, yet only by means of an explicit change of the model formulation. The compromise solutions gained by such modifications are again fuzzy-efficient concearning the reformulated model, but in general not concearning the original model (for a counter-example see [3]). With the modification of the membership function the decision maker is able to approximate any nonlinear membership function locally well by means of the here considered linear membership functions. The results of modifications are ascertained afterwards, thus it is necessary to be able to return to previous solutions.

The main advantage of the flexible mathematical programming system is the relatively simple structure which allows the use of standard software tools and is easily understandable by the decision maker. Some further developments are necessary to integrate efficient compensatory operators into the decision support system.

Literature

[1] R.E. Bellman, L.A. Zadeh, Decision-making in a fuzzy environment, Management Science 17 (1970) B-141 - B-164

[2] S.A. Orlovsky, On programming with fuzzy constraint sets, Kybernetes 6 (1977) 197 - 201

[3] B. Werners, Interaktive Entscheidungsunterstützung durch ein flexibles mathematisches Programmierungssystem (Minerva Publikation, München, 1984)

[4] M. Zeleny, Multiple Criteria Decision-making (McGraw-Hill, New York 1982)

[5] H.-J. Zimmermann, Fuzzy mathematical programming, Computers & Operations Research 10 (1983) 291 - 298

[6] H.-J. Zimmermann, P. Zysno, Decisions and evaluations by hierarchical aggregation of information, Fuzzy Sets and Systems 10 (1983) 243 - 260

AN APPLICATION OF FUZZY OPTIMIZATION TO OPTIMUM ALLOCATION OF PRODUCTION

Jaroslav Ramík [1]

1. INTRODUCTION

In our metallurgical research institute we have encountered the following problem of operative planning: a given daily production of metal of a metallurgical plant should be allocated to individual metal-producing devices being themselves complex systems. This allocation should keep the present-day individual producing characteristics of the devices expressed more or less quantitatively and should ensure minimal cost or maximal profit of the overall production. The mathematical model of such an optimal allocation problem can be formulated as follows:

Maximize

$$z = c_1 x_1^2 + d_1 x_1 + \ldots + c_N x_N^2 + d_N x_N \tag{1}$$

subject to

$$a_{i1} x_1 + a_{i2} x_2 + \ldots + a_{iN} x_N \lesseqqgtr b_i \ , \ i=1,2,\ldots,M \ , \tag{2}$$

$$x_j \gtreqqless 0 \ , \ j=1,2,\ldots,N \ . \tag{3}$$

Simple enough. A quadratic programming problem which can be numerically solved by a classical method of mathematical programming. Here, x_j means the intensity or productivity of the production of the j-th device. The characteristic (nonclassical) feature of the problem is the presence of the parameters c_j, d_j, a_{ij}, b_i which are not strict fixed values, contrary, they are variable, imprecise, frequently vaguely determined characteristics. For planning needs the values of parameters are predicted by experts (engineers for technology). Up till now, the experts are asked to fix the value of a parameter by a definite number and this number is then accepted for computations in mathematical model. Since the mentioned vagueness and variability of parameters is not considered in the model, the obtained solution (decision) is frequently misleading and naturally it is refused by planners (decision-makers).

To increase the quality of decisions made on the basis of the computed mathematical model we have attempted to utilize fuzzy techniques, namely, fuzzy optimization methods. Transforming the problem into the language of fuzzy sets, we meet several aspects which have to be noti-

---

1) Iron and Steel Research Institute, 739 51 Dobrá, Czechoslovakia

ced . The first one is the inequality sign " $\leq$ " in the capacity constaints (2), i.e. there is a problem of ordering or comparing two or more fuzzy values. Since in fuzzy case of parameters c_j , d_j the goal is also fuzzy, we have to decide which element of the family of fuzzy goals is the " best " one. Last but not least, there is a problem of the shape of fuzzy parameters. As indicated by experience it is suitable to restrict the possible amount of fuzzy parameters.

2. THEORETICAL CONSIDERATIONS AND RESULTS

Consider the usual inequation

$$a \leq b ,$$

a , b being real numbers. The most direct way of obtaining the inequality relation between fuzzy values a and b is using the Extension principle, see e.g. [2] , [6] . Doing so we obtain the fuzzy relation between fuzzy numbers a , b as follows:

$$\varrho(a,b) = \sup_{x \leq y} \min \{\mu_a(x), \mu_b(y)\} \quad .$$

As is evident in Fig.1 , such a fuzzy relation doesn't distinguish the fuzzy values with a necessary preciseness. In Fig.1 , there are many values u with $0 = \mu_b(u) < \mu_a(u) = 1$. By definition we have, however, that $\varrho(a,b) = 1$, i.e. a is " less then " b with the membership grade unity.

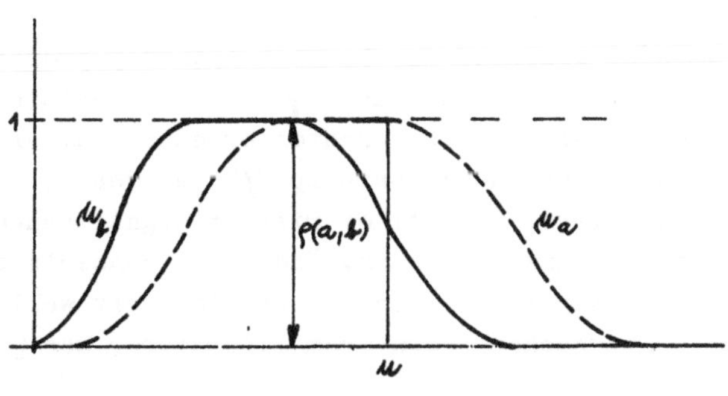

Fig.1

We present a definition of two types of fuzzy preference relation which order fuzzy numbers more delicately and which can be naturally interpreted. Remember that a is a <u>fuzzy number</u> if it is a <u>normal convex fuzzy set</u> in the real line, i.e. there is $u \in E^1$ with $\mu_a(u) = 1$ and for $\tau \in [0,1]$ the <u>τ-level set</u>

$$[a]_\tau = \{ x \in E^1 ; \mu_a(x) \geq \tau \}$$

is a convex strict set in the real line E^1 .

Let a , b be fuzzy numbers, then we define

$$\begin{aligned} \rho_R(a,b) &= 0 \quad \text{if} \quad \sup[a]_1 > \sup[b]_1 , \\ &= 1 - \gamma \quad \text{otherwise} , \end{aligned}$$

where

$$\gamma = \inf\{ \lambda \in [0,1] ; \forall \tau \geq \lambda : \sup[a]_\tau \leq \sup[b]_\tau \}.$$

The fuzzy preference relation ρ_R is called the right inequality relation (shortly, R-relation) . Further,

$$\rho_L(a,b) = \begin{cases} 0 & \text{if} \quad \inf[a]_1 > \inf[b]_1 , \\ 1 - \delta & \text{otherwise} , \end{cases}$$

where

$$\delta = \inf\{ \lambda \in [0,1] ; \forall \tau \geq \lambda : \inf[a]_\tau \leq \inf[b]_\tau \}.$$

The fuzzy preference relation ρ_L is called the left inequality relation (L-relation).

In our optimal allocation problem (1) - (3) we encounter two types of resources, namely, an expended resource $a = a_1x_1+\dots+a_Nx_N$, (or consumption source) on the left hand side of inequality (2) and a resource at disposal b (or capacity resource), on the right side of (2).

The interpretation e.g. of the R-relation is the following. Whatever the value of the expended resource may be with the only property that the possibility of occurrence is at least γ , there exists a sufficient resource at disposal with the greater or equal membership grade, see Fig.2 . In such a case the membership γ serves to a decision-maker as an aspiration level (level of risk or significance level) on which he makes his nonfuzzy decision. The above concepts of fuzzy inequality relation satisfy our allocation problem very well in contrast to other possible approaches known from literature, see [1] , [4] , [5] .

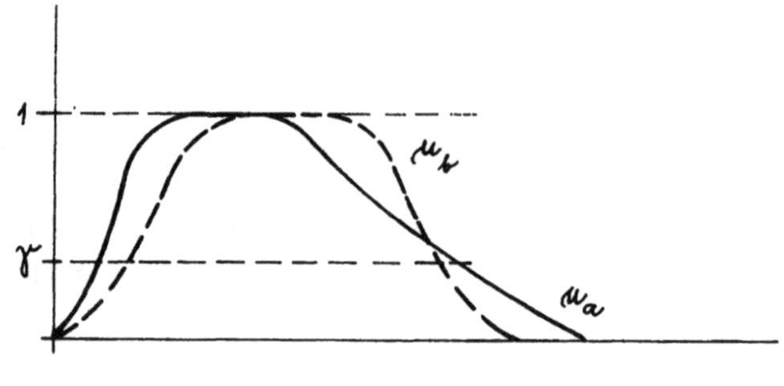

Fig.2

When considering fuzzy parameters in our original model we meet the necessity of deciding which fuzzy goal (i.e. profit or cost) from a given family of goals is the "most rational". We have chosen the approach which is similar to the approach from [3] . First, for a given fuzzy preference relation ϱ we consider the following strong fuzzy preference relation ϱ^s :

$$\varrho^s(a,b) = \max\{ 0 , \varrho(a,b) - \varrho(b,a)\} .$$

For any fixed fuzzy number b the function $\varrho^s(.,b)$ describes a fuzzy set of fuzzy numbers which are strongly determined by b .

Let Z be a set of fuzzy numbers, a fuzzy number $z^* \in Z$ is said to be a ϱ-nondominated element of Z (shortly, ϱ-ND-element) if there is no $z \in Z$ with the property that

$$\varrho^s(z^*,z) > 0 .$$

Taking $\varrho = \varrho_R$ it is possible to demonstrate that if z^* is a ϱ_R-ND-element of Z , then

$$\sup [z^*]_1 \geqq \sup [z]_1 \quad \text{for each } z \in Z . \qquad (4)$$

Conversely, if z^* is a unique element of Z with the property (4) , then z^* is a ϱ_R-ND-element . Similar result can be formulated for $\varrho = \varrho_L$ with "inf" instead of "sup" in (4) . This property is utilized in problem (1) - (3) with trapezoidal fuzzy parameters as will be demonstrated in the following.

As to the type of fuzzy numbers which model the parameters of our problem (1) - (3) , we restrict the broad class of all possible fuzzy numbers to a class of so called trapezoidal fuzzy numbers.

A fuzzy number a with a membership function μ_a is said to be trapezoidal if there are four real numbers $m , n , \alpha , \beta , m \leqq n , \alpha \geqq 0 , \beta \geqq 0$ which form the shape of the membership function μ_a as follows :

$$\begin{aligned}\mu_a(t) &= 1 && \text{for } m \leqq t \leqq n , \\ &= 0 && \text{for } t \leqq m-\alpha \text{ or } t \geqq n+\beta , \\ &= \frac{n+\beta-t}{\beta} && \text{for } n \leqq t \leqq n+\beta , \beta > 0 , \\ &= \frac{t-m+\alpha}{\alpha} && \text{for } m-\alpha \leqq t \leqq m , \alpha > 0 .\end{aligned} \qquad (5)$$

Thus , any trapezoidal fuzzy number a is fully determined by a quadruple of real numbers m , n , α , β as above ; we write

$$a = (m , n , \alpha , \beta) .$$

In Fig.3 different types of trapezoidal fuzzy numbers are depicted.

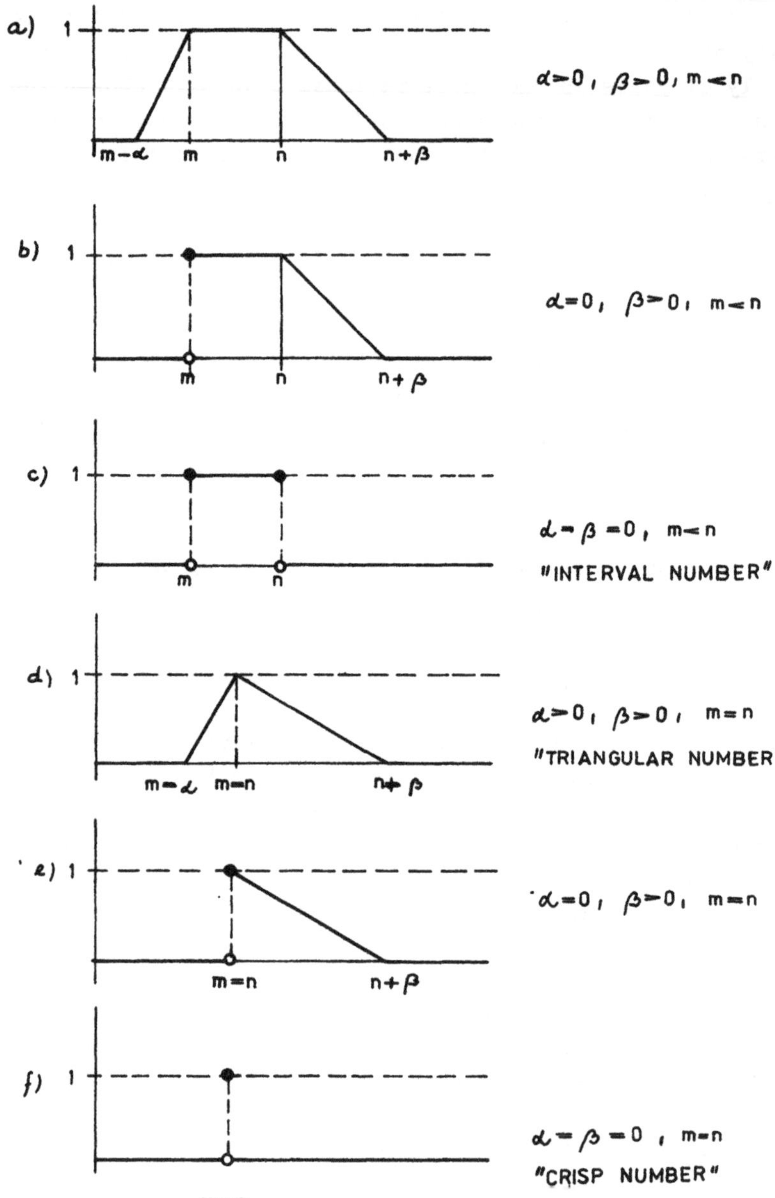

Fig.3

Trapezoidal fuzzy numbers have a remarkable linearity property: for $a_k = (m_k, n_k, \alpha_k, \beta_k)$, $x_k \geqq 0$, $k=1,\dots,p$, the fuzzy number

$$a = a_1x_1 + a_2x_2 + \dots + a_px_p$$

is also trapezoidal such that $a = (m, n, \alpha, \beta)$ with

$$m = \sum m_kx_k , \; n = \sum n_kx_k , \; \alpha = \sum \alpha_kx_k , \; \beta = \sum \beta_kx_k .$$

Here, we use, of course, the Extension principle when doing the linear combination of fuzzy numbers.

In practice, the parameters a_{ij} , b_i , c_j , d_j in (1) - (3) are evaluated by experts who often have not a fully clear idea how to do it. Trapezoidal fuzzy numbers give them simple structure acceptable to human reasoning. Experience indicate that a more complex structure of fuzzy numbers is usually unacceptable by experts. Moreover, trapezoidal fuzzy numbers include crisp numbers , interval numbers, triangular numbers and can be easily handled in mathematical considerations. For trapezoidal fuzzy numbers the introduced fuzzy preference relations ϱ_R and ϱ_L can be expressed as follows:

Take $a = (m_a, n_a, \alpha_a, \beta_a)$, $b = (m_b, n_b, \alpha_b, \beta_b)$ be trapezoidal fuzzy numbers. Then

$$\varrho_R(a,b) = \begin{cases} 0 & \text{if} \quad n_a > n_b \;, \\ \dfrac{n_b - n_a}{\beta_a - \beta_b} & \text{if} \quad n_a \leqq n_b \text{ and } n_a + \beta_a > n_b + \beta_b \;, \\ 1 & \text{othewise} \;. \end{cases} \tag{6}$$

$$\varrho_L(a,b) = \begin{cases} 0 & \text{if} \quad m_a > m_b \;, \\ \dfrac{m_b - m_a}{\alpha_b - \alpha_a} & \text{if} \quad m_a \leqq m_b \text{ and } m_a - \alpha_a > m_b - \alpha_b \;, \\ 1 & \text{otherwise} \;. \end{cases}$$

Now, we investigate the optimum allocation problem (1) -(3) with parameters c_j , d_j , a_{ij} , b_i being trapezoidal fuzzy numbers. Denoting

$$a_i = a_{i1}x_1 + a_{i2}x_2 + \ldots + a_{iN}x_N$$

we can write the membership function μ_i^C of the fuzzy set of alternatives (decisions) satisfying constraint i in the form

$$\mu_i^C(x) = \varrho(a_i, b_i) \;,$$

where $\varrho \in \{\varrho_R, \varrho_L\}$, $i=1,\ldots,M$. To each decision $x \in E^N$ this function designes a degree to which this decision satisfies the constraint i . It is natural to accept that x satisfies all of the constraints to a degree

$$\mu^C(x) = \min_{i=1,\ldots,M} \mu_i^C(x) \;.$$

Here, we accept that the fuzzy set of feasible decisions is the intersection of the fuzzy sets of decisions satisfying the respective constraints.

Choosing an aspiration level τ , $0 < \tau \leqq 1$, a decision-maker obtains a strict set (τ-level set) of feasible decisions which satisfy the constraints with the membership grade greater or equal to τ . By this level set a set of fuzzy goals is determined such that a ϱ-ND-element can be found .

Supposing $\varrho = \varrho_R$, our original problem (1) - (3) is reformulated as follows:

maximize

$$z = \sup \left[c_1 x_1^2 + d_1 x_1 + \ldots + c_N x_N^2 + d_N x_N \right]_1 \tag{7}$$

subject to

$$\varrho_R(a_{11}x_1 + \ldots + a_{1N}x_N \, , \, b_i) \geqq \tau \quad , \; i=1,\ldots,M \, , \tag{8}$$

$$x_j \geqq 0 \quad , \; j=1,\ldots,N \, . \tag{9}$$

Now, let $c_j = (m_{cj}, n_{cj}, \alpha_{cj}, \beta_{cj})$, $d_j = (m_{dj}, n_{dj}, \alpha_{dj}, \beta_{dj})$, $a_{ij} = (m_{aij}, n_{aij}, \alpha_{aij}, \beta_{aij})$, $b_i = (m_{bi}, n_{bi}, \alpha_{bi}, \beta_{bi})$ be given trapezoidal fuzzy parameters. Then by using (6) and (5) we can transform problem (7) - (9) to the following ordinary quadratic programming problem:

maximize

$$z = \sum_{j=1}^{N} (n_{cj} x_j^2 + n_{dj} x_j) \tag{10}$$

subject to

$$\sum_{j=1}^{N} (n_{aij} + \tau \beta_{aij}) x_j \leqq n_{bi} + \tau \beta_{bi} \, , \tag{11}$$

$$\sum_{j=1}^{N} n_{aij} x_j \leqq n_{bi} \quad , \; i=1,\ldots,M, \tag{12}$$

$$x_j \geqq 0 \quad , \; j=1,\ldots,N \, . \tag{13}$$

If $x^* \in E^N$ is a unique optimal solution of (10) - (13) , then $\sum (c_j x_j^{*2} + d_j x_j^*)$ is the ϱ-ND element of the set of all fuzzy goals $\sum (c_j x_j^2 + d_j x_j)$ with x_j satisfying (11) - (13) .

Considering the L-relation ϱ_L an analogical problem to (10) - (13) is formulated:

maximize

$$z = \sum_{j=1}^{N} (m_{cj} x_j^2 + m_{dj} x_j) \tag{14}$$

subject to

$$\sum_{j=1}^{N} (m_{aij} - \tau \alpha_{aij}) x_j \leqq m_{bi} - \tau \alpha_{bi} \, , \tag{15}$$

$$\sum_{j=1}^{N} m_{aij} x_j \leqq m_{bi} \, , \; i=1,\ldots,M \, , \tag{16}$$

$$x_j \geqq 0 \quad , \; j=1,\ldots,N \quad .$$

When solving the original optimal allocation problem a decision-maker chooses a suitable types of fuzzy preference relations ϱ_R or ϱ_L , then successively chooses aspiration levels and for each level the respective optimization problem (10) - (13) or (14) - (17) is solved. The obtained solutions serve to a decision-maker as a basis for making optimal decisions.

3. EXAMPLE

We present here a simple illustrative examle which has, however, some main features of the real application.

Maximize

$$z = c_1x_1^2 + d_1x_1 + c_2x_2^2 + d_2x_2 \ , \tag{18}$$

subject to

$$a_1x_1 + a_2x_2 \leqq b \ , \tag{19}$$

$$x_j \geqq 0 \ , \ j=1,2 \ , \tag{20}$$

c_j , d_j , a_j , b being trapezoidal fuzzy numbers such that

$$\begin{aligned} c_1 &= (-0.3,-0.25,0,0) \ , \\ c_2 &= (-0.5,-0.44,0,0) \ , \\ d_1 &= (0.8,1,0,0) \ , \\ d_2 &= (1,1.33,0,0) \ , \\ a_1 &= (2.5,3,0.3,0.3) \ , \\ a_2 &= (0.8,1,0.3,3) \ , \\ b &= (2,3,1,0.3) \ . \end{aligned} \tag{21}$$

Considering fuzzy relation ϱ_R and aspiration level τ and using (10) - (13), the problem (18) - (21) is reformulated:

maximize

$$z = -0.25x_1^2 + x_1 - 0.44x_2^2 + 1.33x_2$$

subject to

$$\begin{aligned} (3 + 0.3\tau)x_1 + (1 + 3\tau)x_2 &\leqq 3 + 0.3\tau \ , \\ 3x_1 + x_2 &\leqq 3 \ , \\ x_j &\geqq 0 \ , \ j=1,2 \ . \end{aligned}$$

Taking the L-relation ϱ_L' along with (14) - (17) we obtain:

maximize

$$z = -0.3x_1^2 + 0.8x_1 - 0.5x_2^2 + x_2$$

subject to

$$\begin{aligned} (2.5 - 0.3\tau)x_1 + (0.8 - 0.3\tau)x_2 &\leqq 2 - \tau \ , \\ 2.5x_1 + 0.8x_2 &\leqq 2 \ , \\ x_j &\geqq 0 \ , \ j=1,2 \ . \end{aligned}$$

In the table given below results for different values of aspiration level τ and fuzzy relations ϱ_R and ϱ_L are presented.

τ	x_1^*		x_2^*		z	
	φ_L	φ_R	φ_L	φ_R	φ_L	φ_R
0.0	0.26	0.45	0.85	0.45	0.66	0.91
0.1	0.29	0.43	0.85	0.50	0.70	0.94
0.3	0.35	0.39	0.85	0.63	0.73	1.02
0.5	0.40	0.38	0.85	0.78	0.76	1.11
0.8	0.48	0.45	0.85	1.05	0.80	1.31
1.0	0.53	0.59	0.85	1.24	0.83	1.47

REFERENCES

[1] Dubois,D.-Prade,H.:"Ranking Fuzzy Numbers in the Setting of Possibility Theory",Information Sciences 30(1983) 183-224

[2] Orlovski,S.A.:"Problems of Decision Making with Fuzzy Information",Nauka Publishers,Moscow(1981) (in Russian)

[3] Orlovski,S.A.:"Decision Making with a Fuzzy Preference Relation",Fuzzy Sets and Systems 1 (1978) 3,155-167

[4] Orlovski,S.A.:"Mathematical Programming Problems with Fuzzy Parameters",WP-84-xx, International Institute for Applied Systems Analysis,Laxenburg(1984)

[5] Ramík,J.,Římánek,J.:"Inequality Relation Between Fuzzy Numbers and its Use in Fuzzy Optimization",Fuzzy Sets and Systems 16 (1985) 549-564

[6] Zadeh,L.A.:"The Concept of a Linguistic Variable and its Application to Approximate Reasoning",American Elsevier Publ. Comp., New York(1973)

STABILITY IN THE PROBLEM OF MULTICRITERIAL FUZZY DECISION-MAKING

Khurgin J.I. and Polyakov V.V.[1)]

The main drawback of the fuzzy models offered for the description of different real phenomena and processes is an arbitrariness in the determination of the corresponding fuzzy sets membership functions arisen because of the absence of data which makes it possible to determine them unquely.

For instance, to evaluate distribution functions with the help of different statistical methods of hypotheses check-ups based on the experimental data, is possible in statistical models.

Unfortunately when describing fuzzy sets the situation is different, since in principle there are no chances of the objective estimate of the membership function, neither are there methods of hypotheses check-ups with respect to the form of these functions. At best only partial a'priori data which is of general nature usually is known. Most of the papers dedicated to this subject avoid the question of the membership function choice: either fuzzy models are described in supposition that membership functions were somehow determined before or some analytic expression for the membership function is assumed which choice is however not given any proof.

An arbitrariness in the determination of the membership function being "Achilles' heel" of the fuzzy approach restricts possibilities of its correct use. None the less, today there are many successful practical applications of fuzzy models in which membership functions are fixed on the basis of a rather subjective considerations or with the help of some euristic procedures.

The question arises: what is the reason of such paradoxical phenomenon? For this can hardly be explained by lucky coincidences.

Analysing the present situation we have finally come to the conclusion that the precision in the determination of functions was not of significant importance if two conditions held:

1. The choice of initial membership functions is based on the principally correct account of the main tendencies of the phenomenon we describe.
2. Fuzzy model is adequate meaning firstly that initial model grounds are true, and secondary, the way of obtaining of initial data from

---

1) Moscow Institute of Gas- and Oil-Industry (Gubkin-Institute), Moscow USSR

an expert corresponds to its real possibilities.

If these conditions take place the methodologically different effective way of solving this problem arises with which instead of searching the arguments for the choice of the form of the membership function unknown, to define (at least parametrically) its approximate form on the basis of some a'priori data with further evaluation of the result varying with the form of the membership function within some permissible range is enough.

Note, this approach doesn't claim the comprehensive solution of the problem: each concrete situation will certainly cause additional questions.

Thus, we draw to the necessity of the research of the stability of the fuzzy models to the deviations of membership functions which may be regarded as the model parameters. Besides, the problem arises to find and describe the domain of permissible swingings of the membership function, the problem of self-sufficient interest.

Let us proceed to the consideration of the approach described if employed to the problem of the Multicriterial fuzzy decision-making (MFDM). For this problem model stability will mean indepedence or few dependence of the result of the choice on the swinging of functions of membership to fuzzy criteria.

Let us remind the traditional formulation of the MFDM problem. The initial discrete set of variants (or alternatives) X is given, among elements of which we disire to single out, nerally speaking fuzzy subset of the best in certain sense variants D which we'll call fuzzy choice. Variants are evaluated over the totality of heterogeneous criterions $C_1,\dots,C_n$ which are considered to be fuzzy, that is each criterion C_j, $j=\overline{1,n}$, singles out on the set X some fuzzy subset X_j of the best according to this criterion variants, with the corresponding membership function $\mu_j(x)$, $x \in X$, /1/.

For solving this problem the technique based on the construction of the fuzzy binary relationship system (P,R) was offered above /2/. Here R denotes the relationship of not strong preference satisfying reflexivity (xRx) or coherence $\forall$ x,y either xRy, or yRx) properties and P-a relationship of the strong preference defined by antireflexivity ($x\bar{P}x$) and asymmetry ($xPy \Rightarrow y\bar{P}x$) properties.

Membership functions of the corresponding fuzzy relationships are diven at Cartesian product XxX as follows:

$$\mu_R(x,y) = 1 - \frac{1}{W(\pi)} \sum_{K=1}^{n} \pi_K \left[\mu_K(y) \ominus \mu_K(x)\right], \tag{1}$$

$$\mu_P(x,y) = \mu_R(x,y) \ominus \mu_R(y,x), \quad x,y \in X, \tag{2}$$

where π_k - the weight of the criterion C_k, $k=\overline{1,n}$;

$W(\pi)$ - certain normalizing function ($W(\pi) = \sum_{k=1}^{n} \pi_k$ in particular);

$\ominus$ - operation of restricted subtraction: $\alpha \ominus \beta = \max\{\alpha-\beta, 0\}$.

Function of membership to the fuzzy decision D is given by the expression:

$$\mu_D(x) = \frac{1}{|X \setminus x|} \left\{ \sum_{y \in X, y \neq x} \mu_P(x,y) \ominus \sum_{\substack{y \in X \\ y \neq x}} \mu_P(y,x) \right\}, \quad x \in X \qquad (3)$$

where the sign $|\cdot|$ means set capacity.

Thus, this technique of the choice gives rise to the partial rank-ordering of variants on the basis of which optimum choice of the most preferable variants set D^*, $D^* \subset D$ is carried out. With this unlike the usual formulation subset D^* is interpreted as being fuzzy, i.e. for the final decision-making not the only variant maximizing $\mu_D(x)$, is presented but rather several best variants.

Before formulating the notion of the stability of the optimum choice of the most preferable variants set D^*, we'll introduce the notion of model ε-stability.

We'll assume that the deflection of membership vector-function $\vec{\mu}(x) = (\mu_1(x), \dots, \mu_n(x))$ is additive: $\vec{\tilde{\mu}}(x) = \vec{\mu}(x) + \vec{\delta}(x)$, where $\vec{\delta}(x) = (\delta_1(x), \dots, \delta_n(x))$ restricted vector-function: $\|\vec{\delta}(x)\| < \delta < 1, \forall x \in X$; $\|\vec{\delta}(x)\| = \max_k |\delta_k(x)|$.

Let $\rho(A,B)$ be the Hemmihg's distance between the fuzzy sets A and B from X:

$$\rho(A,B) = \frac{1}{|X|} \sum_{x \in X} |\mu_A(x) - \mu_B(x)|$$

and let D and $\tilde{D}$ be the fuzzy solutions of MFDM problem obtained from the model Φ for the initial and deflexible membership vector-function $\vec{\mu}(x)$ and $\vec{\tilde{\mu}}(x)$ respectively.

<u>Definition:</u> Model Φ is called ε-stable, if

$$\forall \varepsilon > 0 \; \exists \delta, 0 < \delta < 1: \|\vec{\delta}(x)\| < \delta \Rightarrow \rho(D, \tilde{D}) < \varepsilon$$

<u>Theorem.</u> Model Φ MFDM (1)-(3) is ε-stable.

This theorem makes it possible to turn from arbitrary monotonous membership functions to parameterized ones, if the latter fill membership function space densely enough.

In practice we confined ourselves to the 2-parametric family of monotonous functions given by the expression:

$$\varphi(x; p, s) = \exp - \left\{ p \left(\frac{x-a}{b-a} \right) \right\}^s, \quad p, s \geq 1, \qquad (4)$$

where a, b- boundaries of membership function varying from the complete membership to the complete non-membership.

Parameter p characterizes the swing of the curve φ: the smaller p is the farther to the right the curve is moved (Fig. 1,a). Parameter s describes the curve's behaviour at the fronts (close to points a,b), i.e. it characterizes the steepness of the curve; with increasing s so increases maximum of the derivative (Fig. 1,b).

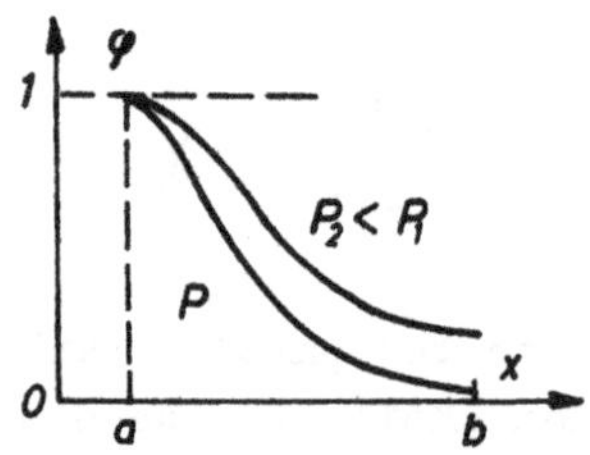

Figure 1,a

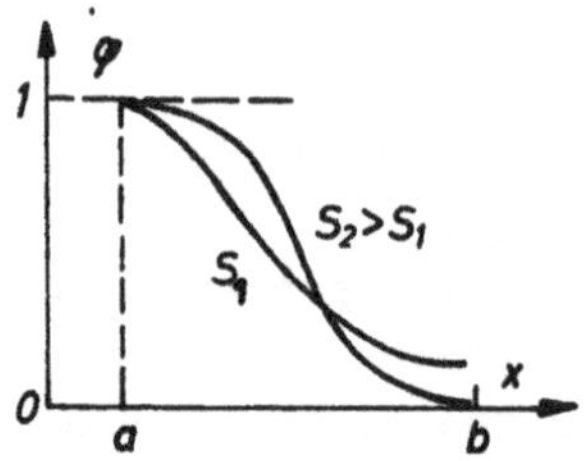

Figure 1,b

The corresponding permissible domain Ω of membership function fluctuations has the approximate form as shown at the Fig. 2.

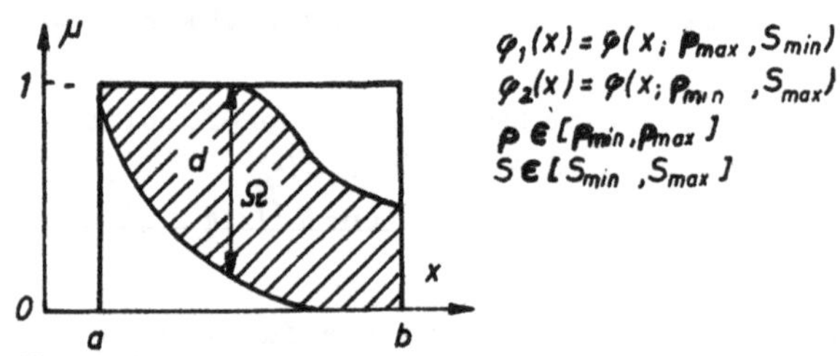

Figure 2

Now let us discuss the stable choice concept. We'll call a width d of the permissible domain the value (Fig. 2):

$$d = \sup_{[a,b]} |\varphi_2(x) - \varphi_1(x)|$$

So far as the multicriterial choice is being discussed it is natural to consider the Cartesian product of domains Ω_i of its component variations $\mu_i(x)$, $i=\overline{1,n}$: $\Omega = \Omega_1 x \ldots x\, \Omega_n$ as the permissible domain Ω for the vector-function $\vec{\mu}(x)$. The corresponding width of the domain Ω is defined by $d = \min_{1 \leq i \leq n} d_i$.

For any membership function lying entirely inside Ω of the width d the most preferable variants set D being the same, we'll call the choice of the most preferable variants set d-stable.

That is the very time for solving in certain sense the reverse problem of finding the permissible variations domains of membership functions corresponding to each stable choice. Then the decision standingfor the maximum domain of permissible variations should be given preference. It is cleare, such a decision reflects the reality to a great extent. However reasons of such situation to have occured, may be deep enough, therefore to test the result obtained with the help of other methods as desirable.

The logics of the reasoning given above, being our ground, we can present a rather general algorithmic scheme of finding of the solutions of the MFDM problem.

First of all consider the rule which is generally speaking not unique of picking out the most preferable variants. Let us show some possible ways:

1. The subset D^* of the most preferable variants is defined by the inequality: $x \in D^*$, if $\mu_D(x) > 1-h$, $0 < h < 1$, where the value of h is given beforehand.
2. $N^* = |D^*|$ - the number of elements of D^* is given, that is N^* elements $x_1, \dots, x_{N^*}$ with the maximum values of the function μ_D are included into D^*.
3. The analysis of the set D is carried out which is based on the assumption that the set D has the certain clustering structure in the sense that the best variants must have rather high and close degrees of the membership to the set D. To find such natural lamination different cluster methods may be used.

Assuming that the rule of finding of the most preferable variants subset D^* is given we'll proceed directly to the consideration of the procedure of stable decisions definition. Its essence is the following. As the result of functions of membership to the fuzzy criteria varying in the permissible range we get the great enough number L of different fuzzy decisions D_l, $l=\overline{1,L}$.

The resulting stable decision D^{st} is defined on the basis of the analysis of these "private" decisions D_l. The idea is that some variants being found all the time among "the best" may be considered to be the best in the global sense, since they are weakly dependent on the possible variability of membership functions. This natural reasoning can be formalized as follows.

Let $\hat{D} = \bigcap_l D_l$ and $\check{D} = \bigcup_l D_l^*$, where $D_l^* \subset D_l$ - the optimum fuzzy decision.

Consider two possible cases:

1. $\hat{D} \neq \emptyset$. This is the most favourable situation. Obviously, $D^{st} = \hat{D}$. With this the corresponding membership function can be defined either by the classic method via minimum:

$$\mu_{D^{st}}(x) = \min_{1 \le l \le L} \mu_{D_l}(x), \quad x \in \hat{D}, \tag{5}$$

or by the less rigid one - via striking an average:

$$\mu_{D^{st}}(x) = \frac{1}{L} \sum_{l=1}^{L} \mu_{D_l}(x), \quad x \in \hat{D} \tag{6}$$

2. $\hat{D} = \emptyset$. In this case we'll define the function of membership to the fuzzy set $\check{D}$:

$$\mu_{D^{st}}(x) = \frac{1}{L} \sum_{l=1}^{L} m_{x,l} \cdot \mu_{D_l^*}(x) \, , \quad x \in \check{D} \tag{7}$$

where

$$m_{x,l} = \begin{cases} 1 \, , & \text{if } x \in D_l^* \\ 0 \, , & \text{in reverse} \end{cases}$$

Then D^{st} is sought as the fuzzy subset of the set $\check{D}$ by means of picking out variants with high values of the function (7). For this the same procedure as for the definition of the subset D_l^*, is used.

Note that in the deterministic interpretation, i.e. D^{st} being considered to be the set in the conventional meaning its definition is trivial:

$$D^{st} = \{ x \in \check{D} \mid \mu_{\check{D}}(x) = \max \mu_{\check{D}}(y) \}$$

The approach described was realized in practice for solving the problem of the choice of the transmission oils best samples. 22 types of oil were presented as the starting material, each type was being evaluated on 18 criteria.

Initial membership functions were being chosen from the curve family (4) and varied independent of each other in the following range of their parameters:

$$p \in [1, 10] \, , \quad s \in [1, 6]$$

The model (1)-(3) was used for solving this problem. With the help of pseudorandom number generator different decisions D_l were being found at this the number of decisions L was being taken equal to 50 and 100. Time for the programme to go - within 15 - 20 seconds. Results showed the stable decision was singled out over the entire permissible domain of membership function swingings which consisted of 6 variants.

The decision obtained coincided with that obtained by the other method (via desirability function /3/) which gave additional certainty in correctness of the decision.

References

/1/ Bellman, R.E., Zadeh, L.A.: Decision-Making in Fuzzy Environment, Management Science, 1970, 17, No. 4, pp. 141-164

/2/ Churgin, J.I., Pojakow, W.W.: Mehrkriterielle Auswahl unter unscharfen Bedingungen und ihre Anwendung bei der Schaffung neuer Erdölprodukte. - In Material zur Vorlesung "Kennwertermittlung und Modellbildung", 1982, 5, S. 5-18.

/3/ Kalinina, E.W., Lapiga, A.G., Churgin, J.I.: Untersuchung der statistischen Eigenschaften der Nutzensfunktion und ihre Anwendung in der Regressionsanalyse - at the same place, S. 26-43

FUZZY ANALYSIS OF THE GROUP CONCORDANCE OF EXPERT PREFERENCES, DEFINED BY SAATY MATRICES

Khurgin J.I. and Polyakov V.V.[1]

Let a problem of objects finite set $C=(C_1,\dots,C_n)$ rank-ordering being solved through definition of some number π_i for each object C_i so that $\sum_{i=1}^{n} \pi_i = 1$, $\pi_i \geq 0$. Vector $\bar{\pi} = (\pi_1,\dots, \pi_n)$ will define rank-ordering required of objects according to their relative importance. In supposition that π_i cannot be measured directly the problem given is solved by means of expert evaluation by the group of m experts. With this each expert carries out pairwise comparison of objects from C and expresses his opinion for every pair with the aid of numerical estimate showing the degree of the importance (or preference) of one object over the other. As the result expert preferences are represented in the form of Saaty matrices which elements are tied with the relationship:

$$a_{ij} \; a_{ji} = 1, \qquad i,j = \overline{1,n} \tag{1}$$

The first employment of such matrices in the expert evaluation problem belongs to Thomas L. Saaty. In /1,2/ he described and substantiated the spectral method of finding the vector $\bar{\pi} = (\pi_1,\dots, \pi_n)$. The method of the definition of the weighting indeces vector over the totality of pairwise comparison matrices itself is not the theme of our discussin, but rather the question whether such solution can be considered reasonable: up to what extent expert preferences on which base we obtain the resulting vector $\bar{\pi}$, are uncontradictory and concordant.

The analysis of the group concordance of expert opinions is generally speaking one of the main problems in expert evaluation methods, as it is obvious enough that with unconcordant estimates final "average" result over the group will be merely unpithy. It means that individual expert preferences in the group being sufficiently concordant one can strike an average by either method.

Currently, a number of techniques of the concordance definition using calculation of different concordance indices of Kendall's indices type or entropy type has been described in literature /3,4/.

Within our problem Kendall's indices appear to be unsuitable, since they are based upon rank-orderings concordance check-up or traditional nonmetric relations. Note, though rankings defined by vectors $\bar{\pi}$ can

[1] Moscow Institute of Gas- and Oil-Industry (Gubkin-Institute), Moscow USSR

coincide, the corresponding components can differ radically. In this sense entropy measures are of little difference from the concordance index.

On account of this fact another index of the group preference concordance based upon fuzzy set theory concepts is offered. Such approach appears to be more adequate to the subjective nature of expert data.

Let A_q be Saaty matrix corresponding to q-th expert, $q = \overline{1,m}$. Obviously, the closer the values of the corresponding matrices A_q elements a_{ij}^q are, the more concordant the result will be. With this it's enough to evaluate the affinity of only the elements of the top matrix triangle which indices belong to the set $J=((i,j):i < j, i=\overline{1,n-1}, j=\overline{2,n})$.

We introduce a notion of concordance of vectors $a_{ij}=(a_{ij}^1,\dots,a_{ij}^m)$, $(i,j) \in J$. With the equality of this vector components it will be considered to be completely concordant. Thus, the notion of the vector concordance can be considered to be fuzzy, so we may introduce the fuzzy set S_m of the concordant vectors $x=(x^1,\dots,x^m)$ with the membership function $\mu_{Sm}(\bar{x})$.

We'll define function μ_{Sm} form constructively as follows. First of all we'll define the notion of the vector concordant in robust sense. Vector $\bar{a}_{ij}$ its components a_{ij}^q being expert estimates of multiplicative nature can be considered to be robust concordant if all or majourity of its components are greater (or less) than 1. Therefore let us introduce the fuzzy set U_m of robust concordant vectors with the membership function as follows:

$$\mu_{Um}(\bar{x}) = 1 - t \, , \tag{2}$$

$$t = \min\left(\frac{m_1}{m-m_1} \, , \, \frac{m_2}{m-m_2}\right) \tag{3}$$

where m_1- the number of vector $\bar{x}$ components less than 1;

m_2- the number of reverse components (greater than 1), $m_1+m_2 \le m$.

With the increasing number of "contradictory" components function (2) converges to zero (Fig. 1).

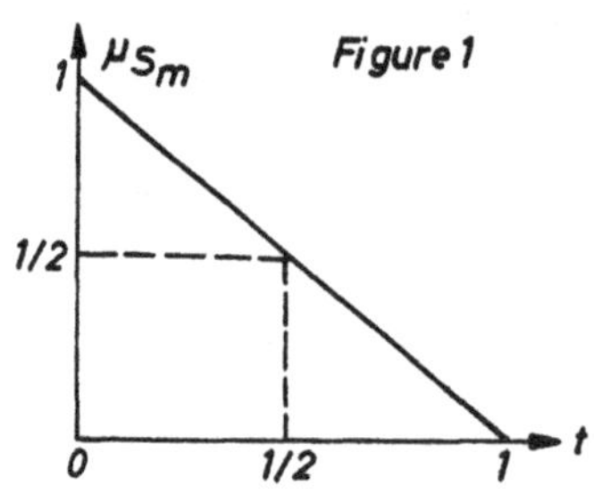

Figure 1

The degree of a vector membership to U_m will be then taken into consideration when constructing function $\mu_{Sm}(\bar{x})$ which is based upon the notion of components deviation with respect to some average value. Due to the multiplicity of Saaty matrix (1) it is natural to consider the following relative quantity as such measure:

$$x^q/G(\bar{x}), \quad q=\overline{1,m}; \qquad G(x) = (\prod_{k=1}^{m} x^k)^{1/m} \tag{4}$$

Then it is natural to take the total deviation of all components as the concordance index of expert estimates:

$$\mu_{Sm}(\bar{x}) = \left[\frac{1}{m} \sum_{q=1}^{m} \Psi\ (x^q/G(\bar{x})) \right]^p \tag{5}$$

$$\Psi(z) = \begin{cases} z, & z \in [0,1] \\ 1/z, & \text{in the reverse} \end{cases}$$

Exponent $p \in [1/2,\ 2]$ depends on the quantity of $\mu_{Um}(\bar{x})$ and from the pithy point of view it "fines" in robust sense badly concordant vector ($p \to 2$) and "encourages" those that are concordant enough ($p \to 1/2$). Exponent p can be determined, for example, by one of the following formulae:

$$1)\ p(u)= \begin{cases} 1/2, & u=1 \\ 1, & u \in [1/2,\ 1) \\ 2, & u \in [0,\ 1/2) \end{cases} \tag{6}$$

$$2)\ p(u)= 2 - 1.5\ u \tag{7}$$

$$3)\ p(u)= u^2+2.5\ u + 2 \tag{8}$$

where $u = \mu_{Um}(\bar{x})$.

Graphs of the expressions (6) - (8) are represented at Fig. 2.

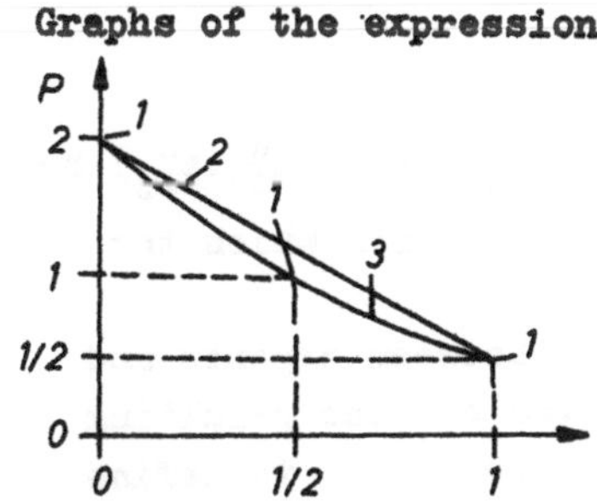

1 - formula (6)

2 - formula (7)

3 - formula (8)

Figure 2

Finding for every pair $(i,j) \in J$ concordance degree $\mu_{Sm}(a_{ij})$ we obtain matrix $F_m = (f_{ij})_n^n$ where $f_{ij} = \mu_{Sm}(\bar{a}_{ij})$ and $f_{ij}' = f_{ji}$.

By calculating arithmetic means of matrix F_m rows we obtain degree of expert concordance with respect to each object from C which can also be useful in further analysis and decision - making.

By striking an average over all elements of matrix F_m we obtain the index of the expert group concordance μ_E:

$$\mu_E = \frac{2}{n(n-1)} \sum_{i=1}^{n-1} \sum_{j=i+1}^{n} f_{ij} \quad (9)$$

The closer to 1 the quantity μ_E is, the more concordant group opinions are. Therefore, some concordance threshold $\beta \in (0,1)$ being given the group is considered to be concordant enough if the following condition holds:

$$\mu_E \geqq \beta \quad (10)$$

If the condition (10) doesn't hold it is necessary to try to find out expert subgroups which concordance is great enough. For this purpose it is necessary to make clustering of expert group with the help of one or several procedures of cluster - analysis /5/.

We'll take the matrix of pairwise experts concordance as the matrix of expert affinity $W = (w_{qs})_m^m$; $w_{qs} = \mu_E(q,s)$, where $\mu_E(q,s)$ - concordance degree of $q^{\underline{th}}$ and $s^{\underline{th}}$ experts, $q,s = \overline{1,m}$. Function $\mu_{S_2}(\bar{x})$ is calculated by formula (5) but some simplier since the fuzzy set U_m with m = 2 degenerates into the crisp one:

$$\mu_{U_2}(\bar{x}) = \begin{cases} 1, & \text{if } t = 0 \\ 0, & \text{if } t = 1 \end{cases}$$

With this exponent p will get only two meanings: 2 and 1/2. Then function (5) will have the following form:

$$\mu_{S_2}(x) = \Upsilon^{p}(x^q/x^s), \quad (11)$$

where $\bar{x} = (x^q, x^s)$.

Having calculated membership degrees $\mu_{S_2}(\bar{a}_{ij})$, $\bar{a}_{ij} = (a_{ij}^q, a_{ij}^s)$ by formula (11) for all $(i,j) \in J$ we obtain matrix F_2 from which the quantity $\mu_E(q,s)$ is defined by formula (9).

Having constructed the matrix W of affinity between experts and making clustering on its base, one ought to find out the resulting opinion about rank-ordering of initial objects set (i.e. to define $\bar{\pi}$) in every sufficiently representative expert subgroup.

Final decision about the set C rank-ordering can be adopted by decision-making person on the base of additional reasoning. For example, vector $\bar{\pi}$ is chosen corresponding to the subgroup of maximum experts.

If one doesn't manage to select representative clusters it is necessary either revise the expedience of this group participation in the examination or make a second round of questionning of Delphy procedure's type /6/.

References

/1/ Saaty, T.L.: A Scaling Method for Priorities in Hierarchical Structures. - J. of Mathematical Psychology, 1977, vol. 15, pp. 234-281.

/2/ Saaty, T.L.: Exploring the Interface between Hierarchies, Multiple Objects and Fuzzy Sets. - J. of Fuzzy Sets and Systems, 1978, vol. 1, pp. 57-68.

/3/ Kendall, M.G.: Rank Correlation Methods, 4th ed. Griffin, London, 1970.

/4/ Евланов, Л.Г., Кутузов, В.А.: Экспертные оценки в управлении. Москва, Экономика 1978.

/5/ Браверман, Э.М., Мучник, И.Б.: Структурные методы обработки эмпирических данных. Москва, Наука 1983.

/6/ The Delphy method. Techniques and Applications. - Reading (Mass.), Addison - Wesley, London, 1975.

FUZZY APPROACH TO THE ANALYSIS OF EXPERT DATA

Khurgin J.I. and Polyakov V.V.[1)]

1. Introduction

Known approaches to the analysis and processing of expert data based on the conventional types of relations and statistical interpretation of expert estimation are often found inadequate to the essence of the problem given.

Expert methods are mainly used in complex systems with high level of uncertainty. To clear out this uncertainty common binary relations of "yes or no" type are often used, though they are not always possible. More adequate would be estimates showing experts degree of certainty that certain relation between objects holds true.

Expert methods based on the statistical approach assume (Thurstone model /1/, for example) the preference degree to have the stochastic nature: it is the probability that one object is more important than the other. With this, the estimate of object importance is random variable and an expert evaluates its every realization. Such model causes serious distruct: there is no probability distribution in the situation when a man evaluates subjectively possible variants, for these are no experiments to be tested numerously under the same conditions with statisticalstability of different outcomes. When analysing the group opinion of the experts presence of some "correct" opinion which is unknown is the main assumption and each separate expert estimation is the "litered" variant of the "correct" one affected by random factors. Hence an automatic transformation of the notion of random variable expectation derives in the form of striking an average group opinion. Expert group is then privately considered to be homogeneous in a certain sense and the result of processing of estimation is considered to be true because it is "average".

As a matter of fact the situation may occur when experts opinion differ radically and then expert group disintegrates into several subgroup with prevailing opinion in each one. In this case, it's clear, that the stochastic model offered won't give any satisfactory result.

For a human mode of thinking vague fuzzy notions and categories of unstatistical nature are more natural. Therefore the elaborate fuzzy set theory seems to be more adequate for the description and analysis of expert data.

---

1) Moscow Institute of Gas- and Oil-Industry (Gubkin-Institute), Moscow USSR

2. Problem formulation

Now let us consider expert techniques using fuzzy set theory language for solving the problem of ranking the set of objects $C = \{ c_1, \ldots, c_n \}$ by introduction of weighting indices of their relative importance. Such question arise, for example, in multicriterial decision-making when it's necessary to rank criterions according to their importance and define their weights /2/.

An expert is supposed to give quantitative data, expressed in the form of preference degree of one objects with respect to the other. For numerous number of objects such metric relations are useful if represented by pairwise comparison matrices, because direct definition of the weights becomes difficult and methods of direct evaluation and ranking are not effective.

In many cases it is worth using multiplicative metric relations $\mathcal{A}$, which are characterizied by the positive matrix $A = \| a_{ij} \| {}^n_n$, which elements are tied with the following relationship:

$$a_{ij} \cdot a_{ji} = 1, \quad i,j = \overline{1,n} \tag{1}$$

From now on we'll call matrices for which condition (1) holds true, Saaty matrices.

Such relation corresponde to the statement of the following type: "object c_i is a_{ij} times as preferable than c_j".

Saaty matrix is called supertransitive, if the relation is of linear order, i.e. the condition of experts estimation concordance holds true /1/:

$$a_{ij} = a_{ik} \cdot a_{kj}, \quad i,j,k = \overline{1,n} \tag{2}$$

3. Scale of Preferences

Generally speaking, arbitrary positive numbers may fit well as a_{ij} estimates but as practice shows to make estimation easy, we better introduce some ratio scale with several points which serves as "mitigated" variant of quantitative estimate. It's worth noting that points are given numerical sense which enables us to make certain arithmetic operations with them permissible in the given class of scales. It's known, for example, that scale of relations in which operation of finding of arithmetic mean holds /1/, stands for the multiplicative relation $\mathcal{A}$.

Let's proceed to the question of ratio scale of preferences construction at the given set of object pairs CxC. Scale may be represented as the unification of R fuzzy classes of preferences $\Pi_0, \ldots, \Pi_{R-1}$, which corresponds to different intensities of preferences. Here we must note that on the one hand we obtain too robust preferences estimates that is too "qualitative" data for small values of R, but on the other, possibilities of erroneous attribution of object pair to the certain class by an expert increase if R is great enough. Therefore it seems sensible to choose some not very small value of R.

The following division of scale of preference intensities into five fuzzy classes appears to be satisfactory:

$(c_i, c_j) \in \Pi_0$ if c_i is equivalent to c_j (E)

$(c_i, c_j) \in \Pi_1$ if c_i is a little more important than c_j (LI)

$(c_i, c_j) \in \Pi_2$ if c_i is rigidly more important than c_j (RI)

$(c_i, c_j) \in \Pi_3$ if c_i is significantly more important than c_j (SI) (3)

$(c_i, c_j) \in \Pi_4$ if c_i is absolutely more important than c_j (AI)

Our task is one of putting certain number σ_r to conformity with each definite class Π_r, $r=\overline{0,R-1}$, so that the degree of condition (2) fulfilment for the elements of matrix A would be maximum. In other words, quantitative proportionality of the matrix elements must stand for logically true expert estimations, i.e. scale σ must best fit coordinated preferences of an expert.

As it follows from the conclusions given above, for fuzzy classes the relation analogous to the condition (2), must hold true and it's best decribed with the help of fuzzy statements language. For example, "if object C_1 is a little more important than C_2 and C_2 is rigidly more important than C_3 then it follows that C_1 is significantly more important than C_3".

Symbolically it can be written as follows:

$$\Pi_1 \;\&\; \Pi_2 \approxeq \Pi_3 \qquad (4)$$

where $\approxeq$ indicates approximate equality.

These logical expressions may include negations of fuzzy statements. But it can be seen that with transporting of negative conjuctive member to another part of fuzzy equality (4) we get equivalent expression. Therefore class indices can be considered to be positive.

We'll map homomorphically empirical system of fuzzy preferences $\{\Pi_r\}$ which satisfies the relations of the (4) type to some numirical system σ keeping relations of empirical system the same. One can easily satisfy himself that, for example, the scale described as

$$\sigma^r = \alpha^r, \quad r = \overline{0,R-1}, \quad \alpha > 1 \qquad (5)$$

and defined up to the parameter precision will fit as such system.

Thechoice of the concrete value of α, is carried out on the base of additional reasons. It is clear, that α depends on the pithy sense which an expert puts in the notion "one object is a little more important than the other". In this situation it seems sensible to assume $\alpha \in (1,2)$.

4. Concordance Analysis of Saaty matrices

So, suppose matrix A is given. To obtain vector $\vec{\pi} = (\pi_1, \ldots, \pi_n)$ of weighting indices of relative importance of several objects $C_1, \ldots, C_n$, spectral method may be used /3/. It aims to find the right eigenvector of matrix A , corresponding to the maximal eigenvalue λ_{max}. If condition (2) doesn't hold true then $\lambda_{max} > n$ and its value is the greater, the less concordance estimation of certain expert are. It's cleare, in case of radical difference between λ_{max} and n there are no reasonable grounds to use the vector obtained as the weighting vector, so we ought to either refuse the data obtained or ask an expert to withdraw his opinion and ponder upon it meticulously.

Before turning to the aggregation of estimates obtained from a group of experts, one has to carry out preliminary evaluation of preferences for each expert. The problem of choosing the criterion for the concordance check-up while comparing values of λ_{max} and n , is poorly developed.

Therefore we can offer another criterion of concordance which appears to be more grounded. Note, that on account of the fact that expert estimations are fuzzy and approximate equality (2) will be. Therefore it seems sensible to use function of membership to some fuzzy set S indicating the degree of the fulfilment of equality (2) as the concordance measure.

It seems this method requires enormous calculation process for it is necessary to make n(n-1)(n-2) check-ups of condition (2), though in fact its size can be lessened and the latter derives from the following assertion.

Assertion: To define the concordance measure of matrix A the check-up of condition (2) fulfilment is necessary for only the elements which indices belong to the set given as follows:

$$J^+ = \{(i,j) : i+1 < j,\ i = \overline{1, n-2};\ \ j = \overline{3,n}\}$$

This requires n(n-1) (n-2)/6 check-ups.

Let us proceed to the definition of function of membership to some fuzzy set $S \triangleq \{x \approxeq y\},\ x, y > 0$, which corresponds to fuzzy fulfilment of the equality between left and right sides of the expression (2).

Set S may be represented in equivalent form: $S \hat{=} \{ Z \approx 1 \}$, where $Z = \max(x,y)/\min(x,y)$. For the fuzzy set given it's necessary to determine the corresponding function $\mu_s(Z)$. Usually the concrete form of membership function is determined apriori, judging from some empirical data or some theoretic reasons. Thus, membership function determination is to certain extent subjective but to consider all the peculiarities of the definite problem will create little distortion to the final result.

Function $\mu_s(Z) = \left[1+((Z-1)/c)^d \right]^{-1}$, $Z \geq 1$, (6)

where c - parameter defining width of function: $\mu_s(c+1)$ 0.5;

d - defines steepnees of curve; the greater the value of d is, the steeper the curve is (Fig. 1).

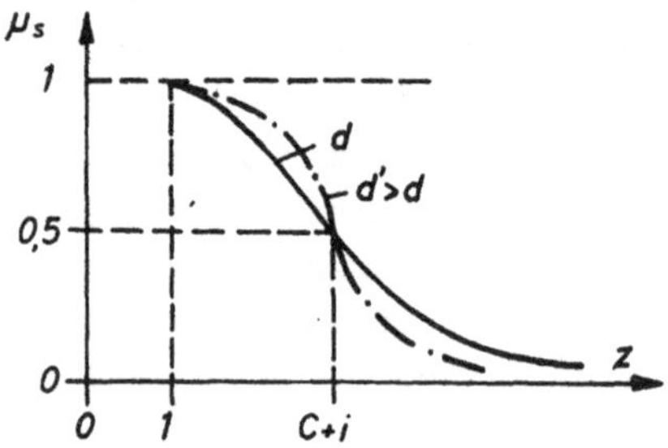

Figure 1

One ought to select parameters c and d varying with the rigidity of requests, laid on the concordance of expert estimations; the more rigid requests are the less value of c and the greater value of d should be chosen.

For example, the following value obtained on the basis of solution of some series of practical problem may be recommended: $c = \sqrt{2}-1$, $d=3$.

Using function (6) offered for discovery of the degree of equalities (2) fulfilment, we can define the total measure of the concordance of certain expert preferences. Arithmetic mean of corresponding values of membership function seems natural to be taken as such measure for each relation (2) on check:

$$\mu_s(A) = \frac{1}{L} \sum_{\ell=1}^{L} \mu_s(z_\ell) , \tag{7}$$

where $L = n(n-1)(n-2)/6$; $z_\ell = \max\{a_{ij}, a_{ik} \cdot a_{kj}\} / \min\{\cdot,\cdot\}$;

$$\ell = (k-i) + (i-1)(j-i-1) + \sum_{s=0}^{j-i-2} s(n-s-1) ,$$

$$(i,j) \in \mathcal{J}^+;\ (i,k),(j,k) \in \mathcal{J},\quad \mathcal{J} = \{(i,k) : i < k\}$$

It's cleare, if matrix is supertransitive $\mu_s(A) = 1$.

Then to make a decision about concordance or non-concordance of pairwise comparison matrix we have to introduce some threshold value $\varsigma \in (0,1)$ so that if $\mu_s(A) > \varsigma$ the matrix is admitted concordant. To choose ς depended upon the requests presented to competence of definite group of specialist-expert and becomes independent problem.

Let us draw our attention to the technique of the discovery of weight vector. With enough concordance it is, generally speaking, not necessary to find eigenvalue λ_{max}. To take mean over each row of matrix preliminarily normalizing elements over columns as the components of vector $\vec{\pi}$ will be enough. Really, for the supertransitive Saaty matrix any of its normalized columns serves as the weighting one. Then it follows if inquality $\mu_s(A) > \varsigma$ holds true, each matrix column will serve as a good enough approximation to its eigenvalue. Striking average over columns elements we must obtain still more precise approximation which as practice showed, almost coincides with the corresponding eigenvector.

Note, with the identical results the method described, is from the points of view of calculation is much simplier than the spectral method.

5. Improvement of concordance of expert preferences

Suppose that after determination of concordance measure some experts found it low, close to the threshold value ς. That means that either considerable number of equality (2) proved to be badly non-concordant or there were not many non-concordant equalities so that the total concordance of matrix, though satisfactory, was not high.

It's obvious that in the first situation we can't make use of that expert's estimates, whereas in the second situation the question arises: why do several non-concordant estimates appear among preferences which are concordant enough?

Here we naturally have to face the fuzzy way of thinking appropriate to the psychology of an expert's brain though not only that. One of the possible reasons of such anomalies to appear is the employment by an expert often unconsciously some accesory extra factors when comparing different pairs of objects. Therefore it's useful to remove or diminish somehow the influence of such factors on expert estimates, leaving an expert apart, all the more that this way is unlikely to bring us to the desirable result. Certain formal procedure of the improvement of expert estimates concordance is offered.

Let us take out subset T, which belongs to this totality and consists of badly non-concordant equalities. It's clear that any of the three elements included in (2) may cause bad difference between left and right sides of an arbitrary equality from T. That's why one has to check-up each element in order to find out limits of its possible variation for improvment of concordance in the equality so that it wouldn't lead to the aggravation of concordance in other equalities which the element considered belongs to.

Formally the condition of this request fulfilment can be expressed as follows. For the element on check a_{ij}, $(i,j) \in J$, we shall calculate $(n-2)$ products $b_{ij}^{k} = a_{ik} \cdot a_{kj}$, $k=\overline{1,n}$; $k \neq i,j$ and consider ordering row of the values obtained

$$B_{ij} = \{ b_{ij}^{1}, \ldots, b_{ij}^{n-2} : b_{ij}^{k} \leq b_{ij}^{k+1} , k = \overline{1,n-3} \}$$

If all members of this row are found greater (less) than a_{ij}, then it means that an expert provided us with in certain sense decreased (increased) estimate while comparing i-th and j-th objects and hence we can increase (decrease) it in principle up to some value $\tilde{a}_{ij}$. Values of $\tilde{a}_{ij}$ are offered to be determined by one of the following formulae:

$$\tilde{a}_{ij} = \begin{cases} \min\limits_{b_{ij}^{k} \neq a_{ij}} \{ b_{ij}^{k} \}, \quad \text{if} \quad a_{ij} \leq b_{ij}^{k}, & (8) \\ \max \{ b_{ij}^{k} \}, \quad \text{if} \quad a_{ij} \geq b_{ij}^{k}, \; k = \overline{1,n-2} & (9) \\ \text{Med}\ (B_{ij}) & (10) \\ G\ (B_{ij}) & (11) \end{cases}$$

where Med denotes median of the row, G - geometrical mean.

One ought to take into consideration that when using expressions (8) and (9) the situation may occur which won't lead to the improvment of concordance but on the contrary to its aggravation. For example, let a_{ij} coincides with all elements of B_{ij}, except the last one which will be significantly greater (or less) than a_{ij}. Calculation of the new value of $\tilde{a}_{ij}$ by formula (8) or (9) will lead to the improvment of concordance in one equality only, instead it will aggravate the concordance in the rest (n-3) equalities and perhaps to a considerable exent. In this case our minimum will be greater than median (from now on we'll consider for the sake of certainty that elements a_{ij} are to be increased; all the relations will hold true in the reverse by substituting in equalities and formula (8) for the formula (9)) therefore when using these formulae one has to compare them with the median (10) without changing element a_{ij} if this relation doesn't hold.

In case $a_{ij} < b_{ij}^{1}$, the choice of $\tilde{a}_{ij}$ according to (8) will be overinsurant meaning that it secures improvment of concordance in all (n-2) equalities though not essential one. The more perceptible effect stands for the employment of (10) and especially (11) because geometrical mean corresponds to the multiplicative nature of matrix A.

Situation, occuring if one part of row elements is greater than a_{ij} and another is less, means that this element "is not responsible" for bad concordance of the equality considered, and we ought to proceed to the analysis of the element next in turn.

Suppose that at least two elements admit alteration. Which variant of alteration should be taken then? To solve this question we may compare alteration of elements values with corresponding elements in matrices of the experts left. If the value of the new estimate of the element is found close to the mean of corresponding estimates given by the experts left, then such alteration may be considered more reasonable under condition that these estimates are concordant enough with each other. Concordance in the group of experts means in robust sense the following situation, for example: one or the majourity of experts consider one object to be more preferable than the other. If expert opinions differ one needn't take his bearings to the mean of group estimates but rather try to either improve concordance of the equality considered at the expense of other elements or, if impossible, increase the element though not badly using formula (8).

Consider the technique of elements alteration with the help of group estimate. Suppose that the concordance of certain equality from T may be improved at the expense of at least two elements included into this equality. Then three cases are possible:

1. all the group estimates are non-concordant; then the alteration should be carried out by (8) (or (9)).
2. at least one group estimate is concordant; if the element altered calculated by (11) differs radically from the group estimate we must try to improve concordance of the equality at the expense of the other elements according point 1.
3. all the group estimates are concordant; the very element is subjected to alteration which has the altered value closest to the corresponding group estimate; the technique of element alteration, thus described, is based on the tendency to strive for the improvment of concordance at the expense of the least possible alterations of matrix A: we'll call it minimum alteration rule.
   So let a_{ij}^{ℓ} be the matrix element of the l-th expert of the group consisting of (m-1) experts. The ranked set of estimate we'll denote as follows:

$$E_{ij} = \left\{ a_{ij}^{1}, \ldots, a_{ij}^{m-1} \; : \; a_{ij}^{\ell} \leq a_{ij}^{\ell+1} \right\}.$$

Having defined $G(E_{ij})$, we subject to alteration the very element which values $\widehat{a}_{ij}$ and $G(E_{ij})$ are closer. In accordance with the minimum alterations rule the final calculation is carried out by the following formulae:

$$\widetilde{a}_{ij} = \begin{cases} G(B_{ij}), & G(B_{ij}) \leq G(E_{ij}) \\ \left.\begin{array}{l} G(E_{ij}),\ a_{ij} \leq G(E_{ij}) \\ a_{ij}\ ,\ a_{ij} > G(E_{ij}) \end{array}\right\} & ,\ G(E_{ij}) < 1 < G(B_{ij}) \\ \mathrm{Med}\left[G(E_{ij}),\ \mathrm{Med}(B_{ij}),\ G(B_{ij})\right], & \text{in reverse.} \end{cases} \tag{12}$$

Let us make a remark on occassion of the last expression: here the definition of median value is essential rather than definition of minimum, since if $a_{ij} > G(E_{ij})$ this element, though to be increased, should be decreased to lead to the still more strong aggravation of concordance. Using the procedure described for the rest of equalities of the set T, we'll obtain the new matrix A with improved concordance which quantitative measure is calculated by (6) and (7).

To conclude this note that in spite of the fact that such analysis requires much to it to apply, the positive effect obtained compensates the efforts wasted. Besides as practice shows, to alter even one element leads to the significant improvement of concordance in several equalities from T simultaneously.

Comparatively easy algorithmization of this procedure makes it possible to employ effectively computers to the high dimension problem.

References

/1/ Литвак, Б.Г.: Экспертная информация. Методы получения и анализа. Москва, Радио и связь 1982.

/2/ Многокритериальные задачи принятия решений. Москва, Машиностроение 1978.

/3/ Saaty, T.L.: Exploring the Interface between Hierarchies, Multiple Objects and Fuzzy Sets.-J. of Fuzzy Sets and Systems, 1978, vol. 1, pp. 57-68.

FUZZY GOALS AND SETS OF CHOICES IN TWO-PERSON GAMES

S.A. Orlovski[1)]

ABSTRACT

We consider two-person games in which sets of feasible choices of the players are described fuzzily with membership functions, and the goals of the players are also described as fuzzy subsets of desired values of the respective pay-off functions. The formulation of fuzzy games considered in this paper is based on the approach to decision making in a fuzzy environment due to R. Bellman and L. Zadeh. Our assumption is that using this approach the players make decisions about choices of their strategies. This assumption enabled us to reduce fuzzily formulated games to their corresponding traditional nonfuzzy formulations and to apply to their analysis game theoretic principles of maximum guaranteed result due to Y.B. Germeyer and of equilibrium due to Nash. Some properties of the respective solutions and their interrelationships in games with opposite preferences of the players are also outlined in this paper.

1. INTRODUCTION

Game is the term commonly used to describe decision making situations involving a number of decision makers typical of many fields of human activities. The principal feature of a game is that the achievement of a certain goal by a player is determined not only by the actions of that player but also by the actions of other players having generally different goals. Therefore, the analysis of such situations purposed at the determination of the rational behaviour of a player should take into account possible actions of all other players. Major efforts in the development of the mathematical theory of games are directed towards the formulation and analysis of principles of rational behaviour of players in various types of games.

Probably the greatest attention of the researchers in this field is attracted by two principles of rational behaviour: the principle of maximum guaranteed result and the principle based on the concept of equilibrium. Both these principles are considered in this paper for the analysis of two-person games in a fuzzy environment, and, here we will outline them briefly.

According to the principle of maximum guaranteed result, a two-person game is viewed as a decision making problem for one of the players (we will call him player 1). Choices of player 2 are considered as uncertain factors of this problem. This uncertainty can mean, for instance, that at the moment of making a choice player 1 will know not a concrete reaction of player 2, but, only some set of his possible reactions (choices). The more player 1 knows about the goals and the possibilities of player 2, the more "narrow" is that set and, therefore, the less the uncertainty with regard to the behaviour of player 2 from the point of view of player 1.

This principle asserts that the rational way of the evaluation by player 1 of his choices consists in taking into account the worst for him reaction of player 2 from the set of possible reactions. Consequently, the best choice by player 1 is a choice having the best evaluation in the above sense.

1) International Institute for Applied Systems Analysis, A-2361 Laxenburg, Austria. The author is on leave from the Computing Centre of the USSR Academy of Sciences, Moscow.

Applying this principle it is important to take into consideration all information about player 2 that player 1 will have when making his concrete choice, since it is that information which enables him to judge about possible reactions of player 2. This flexibly understood principle of maximum guaranteed result and its applications to various types of games are described and analyzed in Y.B. Germeyer 1978.

Another rationality principle considered in this paper is based on the concept of equilibrium due to Nash. This principle is applicable to cases in which the players can communicate with each other to choose a situation that is mutually acceptable. According to this principle rational choices x_0, y_0 of players 1 and 2, respectively, should possess stability properties in the following sense: a unilateral change of this choice is not beneficial to any of the players. In other words, if, for example, player 1 would choose x instead of x_0, and, player 2 would stick to y_0, player 1 will not obtain a better result. A pair (x_0,y_0) having this stability property is called Nash equilibrium situation.

With all its attractiveness this principle has two drawbacks. First, equilibrium situations are often not stable with respect to simultaneous changes of choices by both players (for instance, both players can obtain better results in some other, not necessarily equilibrium situation). Second, there can be a multitude of various equilibrium situations in a game, and different of them can be preferred by different players. Most "preferable" are games in which there are situations rational with respect to both the above rationality principles simultaneously. Many types of so-called zero-sum two-person games (antagonistic games) fall into this category. In this paper we show that in a certain sense this property is characteristic of some games with opposite goals in a fuzzy environment.

In this paper, we deal with games in which players' goals and sets of choices are formulated as fuzzy subsets of respective spaces. And, in our analysis of them we demonstrate the application of both the principles outlined above. We limit the presentation by examples of only two-person games, however, many of the results obtained can be extended to games with greater number of participants.

2. DESCRIPTION OF THE GAME

We consider players 1 and 2 and assume that their feasible choices (strategies) are described respectively by fuzzy subsets $\mu_1:X \to [0,1]$ and $\mu_2:Y \to [0,1]$ of some basic sets X and Y. This description reflects that in the players' viewpoint, choices differ in their feasibility degrees, and, only some of them may be considered as definitely feasible.

We also assume that two pay-off functions $f_1,f_2:X \times Y \to R^1$ are specified with values interpreted as evaluations by the respective players of situations $(x,y) \in X \times Y$.

Each of the players strives to attain his own fuzzily specified goal. We assume that the goal of player i is described as a fuzzy subset $\overline{\mu}_g^i:R^1 \to [0,1]$ of the number axis R^1. Player i tries to obtain a value of $f_i(x,y)$ belonging to $\overline{\mu}_g^i$ to a higher possible degree. Interpretations of this type of goals can be found in many papers based on the use of the Bellman-Zadeh formulation of decision making problems in a fuzzy environment. Rather than repeating those interpretations here, we should note that the goal considered by a player may happen to be poorly or fully incompatible with his possibilities, i.e. with his set of feasible choices and/or the information that he has about the other player. In the following this question is considered in more detail.

To analyze the behaviour of the players we introduce a more explicit description of the goals in the form of fuzzy subsets of $X \times Y$:

$$\mu_g^i(x,y) = \bar{\mu}_g^i(f_i(x,y)) \ , \ i = 1,2, x,y \in X \times Y \quad . \tag{1}$$

Now we introduce fuzzy subsets D_1 and D_2 of $X \times Y$:

$$\mu_{D1}(x,y) = \min \{\mu_1(x) \ , \ \mu_g^1(x,y)\}$$

$$\mu_{D2}(x,y) = \min \{\mu_2(y) \ , \ \mu_g^2(x,y)\}$$

In other words, fuzzy sets D_i, $i = 1,2$ are intersections of the respective fuzzy sets of choices and of the goals. The meaning of D_1 and D_2 can be explained in the following way. If for example player 1 knew a concrete choice $y \in Y$ by player 2, then he would face a problem of attaining his fuzzy goal $\mu_g^1(x,y)$ by making choices from fuzzy set $\mu_1(x)$. Using the approach due to R. Bellman and L. Zadeh (1970), the fuzzy solution to that problem is defined as $\mu_{D1}(x,y)$. Therefore, fuzzy set $\mu_{D1}(x,y)$ can be considered as a family with respect to parameter y of fuzzy solutions to problems of attaining fuzzy goals $\mu_g^1(x,y)$. Similarly, $\mu_{D2}(x,y)$ can be understood.

The value $\mu_{Di}(x,y)$ is the degree to which the fuzzy goal μ_g^i of player i is attained when the players make their respective choices x and y.

We assume in the following that for each fixed choice of one player the other player chooses a strategy that maximizes the respective function $\mu_{Di}(x,y)$, i.e. a strategy belonging to set D_i, to the maximum degree. With this assumption we can consider our game as a traditional (nonfuzzy) two-person game with X, Y being sets of choices of the players and with $\mu_{D1}(x,y)$, $\mu_{D2}(x,y)$ being their respective pay-off functions.

3. MAXIMUM GUARANTEED RESULTS

Let us assume that player 1 can make his choice and communicate it to player 2. Then the maximum guaranteed result of player 1 (maximum guaranteed degree of attaining the fuzzy goal of player 1) is*.

$$M_1 = \max_{x \in X} \min_{y \in Y(x)} \mu_{D1}(x,y) = \max_{x \in X} \min \{\mu_1(x) \ , \ \min_{y \in Y(x)} \mu_g^1(x,y)\} \quad .$$

The set Y(x) in this equation, that depends on x and is known to player 1, is the set of possibile reactions of player 2 on the choice x by player 1. In this sense the mapping Y(x) reflects the knowledge of player 1 about the interests and possibilities of player 2. If for example, player 1 knows only set Y of player 2 then his fuzzy set of choices has the form:

$$\mu_1^1(x) = \min \{\mu_1(x) \ , \ \min_{y \in Y} \mu_g^1(x,y)\} \quad .$$

If, on the other hand, player 1 has more information about player 2 (i.e. in the case of $Y(x) \subseteq Y$, $x \in X$, then the fuzzy set of his choices becomes:

* For simplicity, we assume in the following that max and min are always attained on the corresponding sets.

$$\mu_1^2(x) = \min \{\mu_1(x) , \min_{y \in Y(x)} \mu_g^1(x,y)\} \quad .$$

As can easily be seen, $\mu_1^2(x) \geq \mu_1^1(x)$, $x \in X$, which means that with more information player 1 has a wider fuzzy set of choices. In this sense, we can say that having more information about his opponent, player 1 has more possibilities to guarantee the attainment of his goal.

If the value of M_1 is too small, then the goal that player 1 tries to attain is too high with respect to his possibilities. In this connection the following problem is of interest: given a formulation of a fuzzy goal of player 1 in the form of a fuzzy subset $\bar{\mu}_g^1$ what should be his fuzzy set of choices that can guarantee the attainment of this goal to a degree not smaller than α? In other words, what should be the function $\mu_1(x)$ describing the set of strategies of player 1, to ensure:

$$M_1 = \max_{x \in X} \min \{\mu_1(x) , \min_{y \in Y(x)} \mu_g^1(x,y)\} \geq \alpha \quad ?$$

If we introduce a set

$$X_\alpha = \{ \min_{y \in Y(x)} \mu_g^1(x,y) \geq \alpha\} \subseteq X \quad ,$$

then it can easily be seen that when $X_\alpha = \emptyset$ we may have $M_1 < \alpha$, and, therefore, in this case player 1 may not be able to guarantee the attainment of his goal to the desired degree whatever his set of strategies is. (The reason, for instance, can be the lack of information about player 2 which is determined by the mapping $Y(x)$).

If $X_\alpha \neq \emptyset$ then $M_1 \geq \alpha$ can be guaranteed if $\mu_1(x) > \alpha$ for at least one $X \in X_\alpha$.

We can also consider an inverse problem: given a fuzzy set of choices of player 1 the attainment of what goals can he guarantee to a degree not smaller than α?

To analyze this problem let us introduce a set:

$$X'_\alpha = \{x \mid \mu_1(x) \geq \alpha\} \quad .$$

A goal $\mu_g^1(x,y)$ can be attained to a degree not smaller than α if

$$\min_{y \in Y(x')} \mu_g^1(x',y) \geq \alpha$$

for some $x' \in X'_\alpha$, or

$$\mu_g^1(x',y) \geq \alpha \ , \quad y \in Y(x) \quad .$$

Let $\bar{\mu}_g^1 : R^1 \to [0,1]$ be some goal function of player 1. We introduce a set

$$Z_\alpha = \{z \mid \bar{\mu}_g^1(z) \geq \alpha\} \subseteq R^1$$

Now we can conclude that a goal $\bar{\mu}_g^1$ can be attained to a degree not smaller than α by player 1, if for at least one $x \in X'_\alpha$ we have $f_1(x,y) \in Z_\alpha$ for any $y \in Y(x)$, where $f_1(x,y)$ is the pay-off function introduced in Section 2.

4. GAMES WITH OPPOSITE PREFERENCES OF THE PLAYERS

If X and Y are the respective sets of choices of the players, then the product set $X \times Y$ is called the set of situations of the game. A situation of the game has the form $s = (x,y)$ with $x \in X$, $y \in Y$. If for player i situation s_1 is more preferable than situation s_2, we shall write $s_1 >^i s_2$.

The difference between the players' preferences in a game can mean, for example, that for some situations s_1, s_2, s_3, and s_4 we have

$$s_1 >^1 s_2 \ , \ s_1 >^2 s_2 \ , \ s_3 >^1 s_4 \ , \ s_4 >^2 s_3 \quad ,$$

i.e. preferences of the players can be the same for some pairs of situations (s_1 and s_2) and opposite for other pairs (s_3 and s_4).

If preferences of the players are opposite *for all situations*, then we have a game with opposite preferences of the players. In other words, in such a game for every pair of situations s and q we have

$$s >^1 q \iff q >^2 s$$

Similarly, by a game with opposite preferences in a fuzzy environment we shall mean a game in which the goal functions $\bar{\mu}^i_g : R^1 \to [0,1]$, $i = 1,2$ and the pay-off functions $f_i : X \times Y \to R^1$ are such that

$$\bar{\mu}^1_g(z_1) > \bar{\mu}^1_g(z_2) \iff \bar{\mu}^2_g(z_1) < \bar{\mu}^2_g(z_2)$$

$$\bar{\mu}^1_g(z_1) = \bar{\mu}^1_g(z_2) \iff \bar{\mu}^2_g(z_1) = \bar{\mu}^2_g(z_2)$$

$$f_1(x,y) = f_2(x,y) \ , \ (x,y) \in X \times Y$$

Using the above definitions (1) of functions $\mu^i_g(x,y)$ (Section 2) we can formulate this definition of a game with opposite preferences in the following more explicit form:

$$\mu^1_g(x_1,y_1) > \mu^1_g(x_2,y_2) \iff \mu^2_g(x_1,y_1) < \mu^2_g(x_2,y_2)$$

$$\mu^1_g(x_1,y_1) = \mu^1_g(x_2,y_2) \iff \mu^2_g(x_1,y_1) = \mu^2_g(x_2,y_2)$$

for any (x_1,y_1), (x_2,y_2).

Similar to Section 2 we shall consider this game as the game with pay-off functions:

$$\mu_{D1}(x,y) = \min\{\mu_1(x) \ , \ \mu^1_g(x,y)\} \ .$$

$$\mu_{D2}(x,y) = \min\{\mu_2(y) \ , \ \mu^2_g(x,y)\}$$

with X and Y being sets of choices of players 1 and 2 respectively.

A pair of choices (x_0,y_0) is called an equilibrium situation if for any $x \in X$ and $y \in Y$ we have

$$\mu_{D1}(x_0,y_0) \geq \mu_{D1}(x,y_0) \qquad (3)$$

$$\mu_{D2}(x_0,y_0) \geq \mu_{D2}(x_0,y) \qquad (4)$$

To consider some interesting properties of this type of equilibrium situations we introduce two sets:

$$C_1 = \{(x,y) \mid (x,y) \in X \times Y \ , \ \mu_1(x) \geq \mu^1_g(x,y)\}$$

$$C_2 = \{(x,y) \mid (x,y) \in X \times Y \ , \ \mu_2(x) \geq \mu^2_g(x,y)\}$$

In the case when an equilibrium situation $(x_0,y_0) \in C_1$ we can conclude that the set

$$C_{1y_0} = \{x \mid x \in X \ , \ (x,y_0) \in C_1\}$$

is not empty. Therefore, as follows from the definition of C_1 we have that $\mu_1(x) \geq \mu_g^1(x,y_0)$ for any $x \in C_{1y_0}$, hence

$$\min \{\mu_1(x), \mu_g^1(x,y_0)\} = \mu_g^1(x,y_0), \quad \text{any } x \in C_{1y_0}$$

But then we have from (3) that $\mu_g^1(x_0,y_0) \geq \mu_g^1(x,y_0)$ for any $x \in C_{1y_0}$.

From this and from definition (2) of the opposite preferences we have $\mu_g^2(x_0,y_0) \leq \mu_g^2(x,y_0)$ for any $x \in C_{1y_0}$, and therefore $\mu_{D2}(x_0,y_0) \leq \mu_{D2}(x,y_0)$. Using (4) we can also write

$$\mu_{D2}(x_0,y_0) \geq \mu_{D2}(x_0,y), \quad y \in C_{1x_0}$$

where

$$C_{1x_0} = \{y \mid y \in Y, (x_0,y) \in C_1\}$$

Finally, we have

$$\mu_{D2}(x_0,y_0) \geq \mu_{D2}(x_0,y), \quad y \in C_{1x_0}$$

$$-\mu_{D2}(x_0,y_0) \geq -\mu_{D2}(x,y_0), \quad x \in C_{1y_0}$$

The last two inequalities mean that the equilibrium situation $(x_0,y_0) \in C_1$ considered is an equilibrium situation of a zero-sum game (μ_{D2} - pay-off function of player 1, and $-\mu_{D2}$ - pay-off function of player 2: $\mu_{D2} + (-\mu_{D2}) = 0$) in which only those situations are feasible which belong to set $C_1 \subseteq X \times Y$ (but not all situations in $X \times Y$). Games of this type are referred to as games with prohibited situations or games with constraints (see, for instance, Orlovski 1973, 1975).

Similarly, it can be shown that in case when an equilibrium situation $(x_0,y_0) \notin C_1 \cup C_2$, then this situation is an equilibrium situation in the game where $\mu_1(x)$ is the pay-off function of player 1, $\mu_2(y)$ - pay-off function of player 2 and the set of feasible situations is $\overline{C}_1 \cap \overline{C}_2$, where $\overline{C}_i = X \times Y \setminus C_i$, $i = 1,2$.

5. MAXIMUM GUARANTEED RESULT AND EQUILIBRIUM SITUATIONS

In a traditional (nonfuzzy) game with opposite interests the players in any equilibrium situation receive their maximum guaranteed pay-offs. In view of this it is of interest to analyze the relation between maximum guaranteed results of the players and their results in equilibrium situations in a fuzzy game with opposite preferences.

In calculating maximum guaranteed results we shall assume that each of the players knows only the basic set of choices of his opponent (X or Y, respectively for players 1 and 2). In such a case the maximum guaranteed results have the following form:

$$M_1 = \max_{x \in X} \min_{y \in Y} \mu_{D1}(x,y)$$

$$M_2 = \max_{y \in Y} \min_{x \in X} \mu_{D2}(x,y)$$

Let us represent M_1 in the form:

$$M_1 = \max \{ \max_{x \in X'_0} \mu_1(x) \,.\, \max_{x \in X_0} \min_{y \in C_{1x}} \mu_g^1(x,y)\} \tag{5}$$

with

$$X_0 = \{x \mid x \in X , C_{1x} \neq \emptyset\} , X_0' = X \setminus X_0 ,$$

$$C_{1x} = \{y \mid y \in Y , (x,y) \in C_1\} ,$$

and C_1 as introduced in Section 3.

For simplicity we assume that $C_1 = C_2 = C$. Let (x_0,y_0) be an equilibrium situation of the game, and $(x_0,y_0) \in C$. In the following we show that in this case the pay-offs of the players in the equilibrium situation (x_0,y_0) are their maximum guaranteed results.

1. To show first that $M_1 \leq \mu_{D_1}(x_0,y_0)$ we note that

$$\min_{y \in Y} \mu_{D_1}(x,y') \leq \mu_{D_1}(x,y_0) ,$$

therefore,

$$M_1 = \max_{x \in X} \min_{y \in Y} \mu_{D_1}(x,y) \leq \max_{x \in X} \mu_{D_1}(x,y_0) = \mu_{D_1}(x_0,y_0)$$

i.e. we obtain the desired inequality.

2. To show that $M_1 \geq \mu_{D_1}(x_0,y_0)$ we note that

$$\mu_g^2(x_0,y_0) \geq \mu_g^2(x_0,y)$$

for any $y \in C_{x_0} = \{y \mid (x_0,y) \in C\}$.

From here and using the definition (2) of opposite preferences of the players (Section 2) we conclude that for any $y \in C_{x_0}$

$$\mu_g^1(x_0,y_0) \leq \mu_g^1(x_0,y)$$

hence

$$\min \{\mu_1(x_0) , \mu_g^1(x_0,y_0)\} \leq \min \{\mu_1(x_0) , \mu_g^1(x_0,y)\} ,$$

and therefore

$$\mu_{D_1}(x_0,y_0) = \min_{y \in C_{x_0}} \mu_{D_1}(x_0,y) = \min_{y \in C_{x_0}} \mu_g^1(x_0,y) .$$

As can easily be seen

$$\max_{x \in X_0} \min_{y \in C_{x_0}} \mu_{D_1}(x,y) \geq \min_{y \in C_{x_0}} \mu_{D_1}(x_0,y) = \mu_{D_1}(x_0,y_0) .$$

From here and using (5) we have

$$M_1 \geq \mu_{D_1}(x_0,y_0)$$

Finally, we have $M_1 = \mu_{D_1}(x_0,y_0)$. Similarly, it can be shown that under the same assumptions we have $M_2 = \mu_{D_2}(x_0,y_0)$.

Now we can conclude that in a game with opposite preferences formulated here the equilibrium degrees of attainment of the players' goals corresponding to an equilibrium situation from set C are equal to maximum guaranteed degrees of their attainment. In particular, this means that all equilibrium situations from set C are equivalent for both the players since in any of them they obtain the same respective degrees M_1 and M_2 of their fuzzy goals.

6. SUMMARY

The formulation of fuzzy games considered in this paper is based on the approach to decision making in a fuzzy environment due to R. Bellman and L. Zadeh. Our assumption is that that using this approach the players make decisions about choices of their strategies. This assumption enabled us to reduce fuzzily formulated games to their corresponding traditional nonfuzzy formulations and to apply to their analyses game theoretic principles of maximum guaranteed result due to Y.B. Germeyer and of equilibrium due to Nash. Some properties of the respective solutions and their interrelationships in games with opposite preferences of the players are also outlined in this paper.

REFERENCES

[1] Bellman, R.E. and L.A. Zadeh: Decision Making in a Fuzzy Environment, *Management Science*, 17 (1970), 141-164.

[2] Germeyer, Y.B.: Games with Nonantagonistic Interests. *Nauka Publ.*, Moscow (in Russian), 1978.

[3] Orlovski, S.A.: Infinite Games with Prohibited Situations. *J. Comput. Math. and Math. Phys.*, No. 13 (1973), 775-781 (in Russian).

[4] Orlovski, S.A.: Equilibrium Situations in Non-Coalitional Games with Constraints. *J. Comput. Math. and Math. Phys.*, No. 15 (1975), 1597-1601 (in Russian).

MODEL BUILDING AND FUZZY SYSTEMS

Bernd Straube [1])

Abstract. Problems encountered in modelling are discussed. In this context the role of fuzziness is regarded. Some of the author's models of real world objects are summerized.

1. Modelling and fuzziness

A well-working model of a real world object is always the result of a successful cooperation between scientists working in the field of system theory and analysis on the one hand and the specialists for the real object on the other.

In general, the claim of the model to be set up will be formulated by the specialists. This claim can impact with the possibilities of the methods used in system sciences and this, in turn, will give impulses for their further development.
The specialists, however, are challenged by system sciences to find out hidden qualitative and quantitative relationships in their object.

Not all of the requirements are able to be fulfilled by the partners. Nevertheless, the model established can be regarded as the best compromise which could be achieved at the moment.

In most cases, the aim consists in finding a model that captures the behavioural aspects of the real object under investigation. Many importent problems to be solved such as prediction, control, classification, pattern recognition, diagnosis, decisions-making, safety analysis, etc. are described as behavioural models.

The following items have to be considered during the process of modelling.

- Which claims and tasks are to be fulfilled by means of the model where a certain cost-effect-relation and a necessary quality of the responces of the model are taken into account ?
- Which of the quantitives acting upon the system are relevant with regard to the intended task of the model ?
- What are the structural relationships between the relevant variables ?
- What are the types of scales of the available data ?
- What methods, techniques, theories, etc. are adequate concerning the problems mentioned before ?

The three questions listed at first consider qualitative features of the process of modelling mainly. The last question should not be taken into consideration too early. For instance, applying a certain method will cause in restrictions of the selection of the relevant variables before treating this point.

1) Akademie der Wissenschaften der DDR,
Zentralinstitut für Kybernetik und Informationsprozesse, Berlin

In general, the treatment of each of the above items will result in starting a new modelling process on this level of abstraction. Answering one of the questions might have an effect on those ones that have already been treated.

Hence, model building is both a hierarchical and 'iterative' process.

The fuzziness of the real world problem has to be included by using elements and ideas of fuzzy set theory. Then a better result in modelling will be achieved since fuzzy phenomena can be modelled.
For instance, each variable that might be relevant to the model can be included in form of a fuzzy variable, at least. Furthermore, heuristic algorithms can often be made more efficiently by means of fuzzy sets.

In describing real objects and phenomena fuzziness may be caused by

- only qualitative knowledge or experience about the object,
- only vague statements are available,
- imprecise formulations of the interactions of some of the compartments of the object,
- using variables with values having no sharp boundaries (such as linguistic ones)
- using variables with a low level of scale.

Up to now, it has been an open problem how to 'map' the fuzziness of the real object to an appropriate mathematical object, i.e. the fuzzy set. This 'mapping' is a model building itself.
There is not available a generally accepted estimation theory for membership functions. Some heuristic strategies for the assessment of the grades of membership are

- aggregation of elementary fuzzy sets and fuzzy subsets under the consideration of disjunctive and conjunctive points of view
- estimation by experts
- use of classes of parametric membership functions and their fitting of their parameters to the real situation

However, the only normalization of a function onto the interval [0,1] must not mix up with the interpretation that the values of the function are thought of as the grades of membership.

As for the fuzzification of a mathematical operation, the extensions principle of ZADEH should be used carefully. It had better model the idea or heuristics this operation is based upon by a fuzzy algorithm. In this respect, one has to realize that the mathematical operation is only another model for it. Therefore, fuzzification could be regarded as modelling of a model.

2. Fuzzy models

As already mentioned, in modelling two aspects have to be considered [6]

- the qualitative or structural aspect
- the quantitative or relational (functional) aspect.

The qualitative interactions between the relevant variables can be described by structural relations. In complex systems it should always be possible to find out internal connexions. This leads to a structured model. The complete model is regarded to be an interconnexion of submodels. Each of the submodels can have a structure again. Hierarchical models possess special kinds of structures.
Finding out a structure important and additional information is obtained about the system and, moreover, for the process of modelling.

A structural relation can also be a fuzzy one. In this case, care should be taken in interpreting the grades of membership.

As soon as a structure has been found each of the submodels can be identified by an appropriate method. Some submodels are derived from a phenomenological point of view, others from experimental process analysis and others, in turn, from a theoretical one. This will yield in an interaction of deterministic, stochastic, and fuzzy submodels.

Thus, the claim of model building to bring in as much knowledge about the object as possible has been paid attention.

Quantitative interrelations are described by relations or functions. There exist different representations of them: an explicitely given computational instruction, an algorithm, a table of corresponding values or sets, a system of equations, etc.

Behavioural models are mostly represented by input-output models. Special input and output variables have to be derived from the relevant variables of the (sub-) model under investigation. A successful way is the consideration of the classification property of an input-output model. Then, the input variables are referred to as the features and the values of the output variable as classes.
Therefore, there is no methodological difference between classification, pattern recognition, situations recognition, diagnosis, decision-making, etc. All of these stress only certain aspects.

The 'right' choise of the features is a crucial point. It has to keep in mind that the feature extraction depends on the relevant information to be filtered.
In Figure 1 a protection of a relation $Q \subset X_1 \times X_2 \times X_3 \times Y$ onto $X_1 \times X_2 \times X_3$ is depicted. Each of the classes K_1, K_2, and K_3 corresponds with an other output value. Because of its (assumed) non-metric level of scale (e.g. linguistical) the feature X_3 has been omitted during the feature extraction process. Then, only a relation $\tilde{Q} \subset X_1 \times X_2 \times Y$ could have been received. But, the classes are no longer disjoint (see Figure 2).

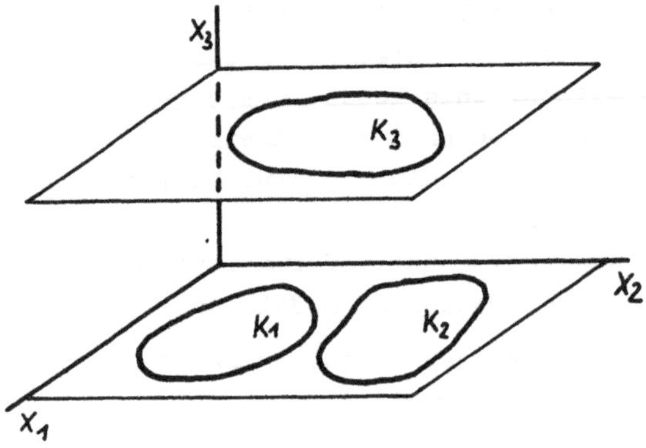

Figure 1: Projection of a relation $Q \subset X_1 \times X_2 \times X_3 \times Y$ onto $X_1 \times X_2 \times X_3$

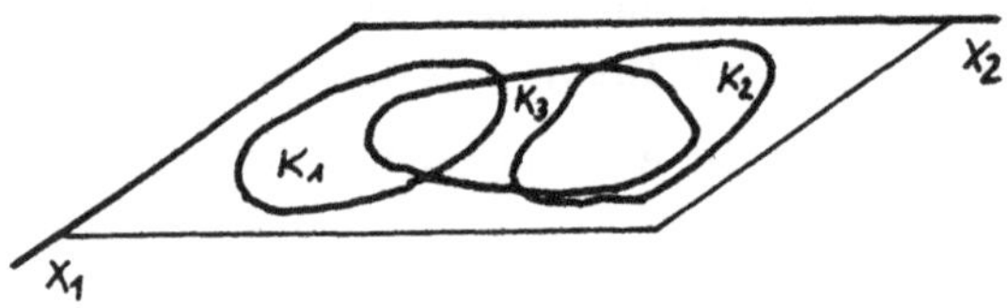

Figure 2: Projection of the relation Q (cf. Fig. 1) onto $X_1 \times X_2$

For fuzzy models fuzzy relations are used. Here, some possiblities for setting up a fuzzy relation are listed.

- Cause-effect relationships are often formulated as a set of 'IF ... THEN'-statements.
  If the causes or effects are expressed by fuzzy sets (e.g. 'large', 'small', 'slightly increasing', ...) they can be described by fuzzy implications. There are different definitions of fuzzy implications.
  A fuzzy 'if...then' statement or fuzzy conditional statement is regarded as fuzzy subrelation. Their fuzzy union constitutes an estimation of the complete fuzzy relation required.
  This set of fuzzy conditional statements can also be regarded as a function that is described by means of corresponding sets.
- Given the corresponding fuzzy inputs and fuzzy outputs the fuzzy relation is determined by solving fuzzy relational equations [5].
- Given measured input-output pairs that are regarded as typical input-output situations of the process under investigation. It is assumed the process behaves similarly in the neighbourhood of a measured point (x_i, y_i). Then, this similarity is expressed by an unimodal fuzzy set (see Figure 3), that is regarded as a fuzzy subrelation again. Their fuzzy union forms the required relation.
- The assumed similarity of the item listed before is used for a cluster analysis of the set of measured input-output pairs. Basing on the clusters an classifier can be derived. It is also possible to attach a parametric model to each cluster. The input-output pairs are used to estimate the parameters. Each model is defined only on its pertinent cluster.
- Sometimes, the process is given by a set of deterministic equations resulting from a theoretical process analysis. However, the parameters are fuzzy since only assumptions or some experience are available. Then, the fuzziness of these parameters is projected onto the solution set. Hence, a fuzzy solution set is obtained. [1]

3. Applications

Basing on the ideas of model building outlined above several applications for real world problems have been accomplished.
The models works in the fields of regional planning, forecast in river systems, diagnosis of electric power switches, electrical network analysis, safety analysis, psycholocial diagnosis in industrial plants, classisfication for several diseases, simulation of ecologigal systems. The numerical results have been obtained by using the FORTRAN IV program package FUZSIM [7] (cf. section 4).
Some of these applications are outlined in the following.

3.1 Fuzzy forecast models for river systems [2],[3]

This problem has been tackled since 1976. Three principal types of models have been investigated.

Model 1
For the corresponding points of time the relevant variables are chosen from the time series of discharge or water level. A set of input-output pairs $P = \bigcup_{i=1}^{n} \{(x_i, y_i)\}$ will be obtained. Typical input-output situations, as characteristic periods of flood, mean water and low water, have to be used in making up the set P. The similar behaviour assumed in the neigbourhood of a pair (x_i, y_i) is described by unimodal fuzzy sets (cf. Figure 3).

Model 2
In opposition to model 1 the set P is clustered fuzzily. The points of a cluster are the basis to identify the parameters of an easy parametric expression.
The model consists of two submodels, a decision model and a model containing a set of parametric equations (cf. Figure 4). The decision model recognizes the situation at the stations. The result is a fuzzy set on the set of indices. Each index corresponds with an equation of the second model. The pertinent equations are chosen in accordance with the two largest membership values of the index set.

Model 3
A structure of coupled subsytems has been found. This model consists of five fuzzy relations based on fuzzy conditional statements (cf. Figure 5). Three of them classify the dynamic hydrological situation for three different stations in a fuzzy manner. One of them classifies the static hydrological situation at the forecast station. The last relation combines the fuzzy results obtained by mentioned submodels with the forecast value.
Not only single forecast values but also short sections of time series are forecasted.

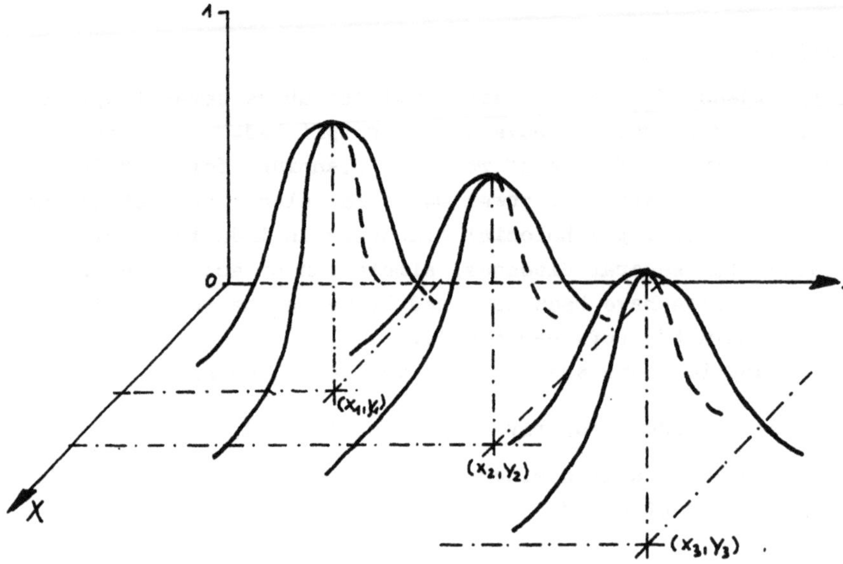

Figure 3: Fuzzy relation based on fuzzified measurements

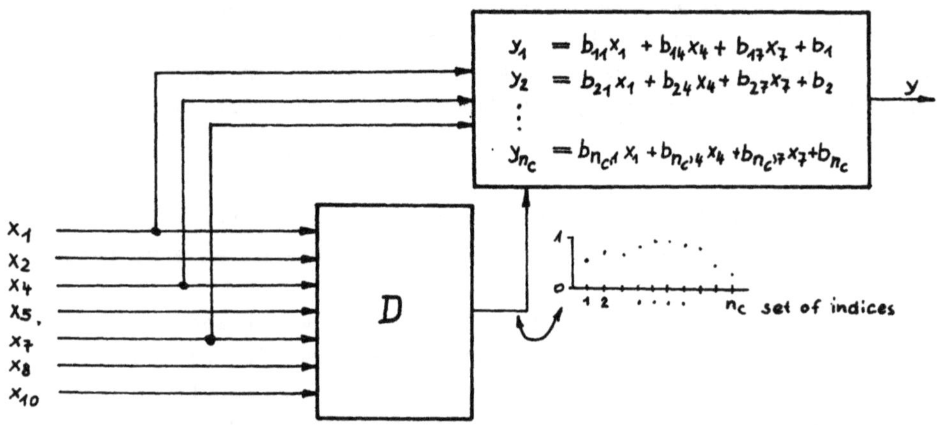

Figure 4: Forecast model with situation recognition submodel D and a set of subscripted parametric models

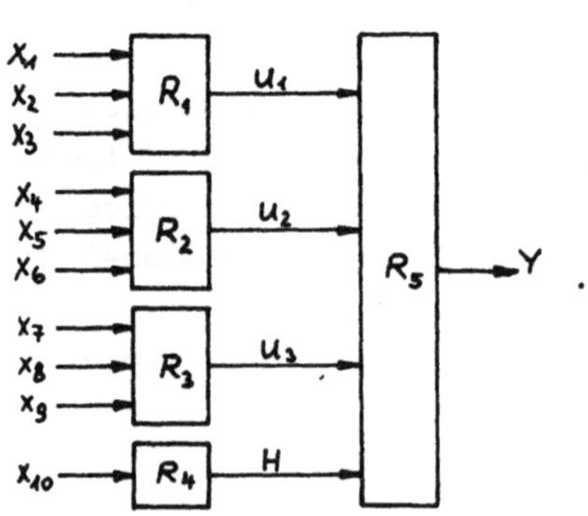

Figure 5: Structured fuzzy forecast model

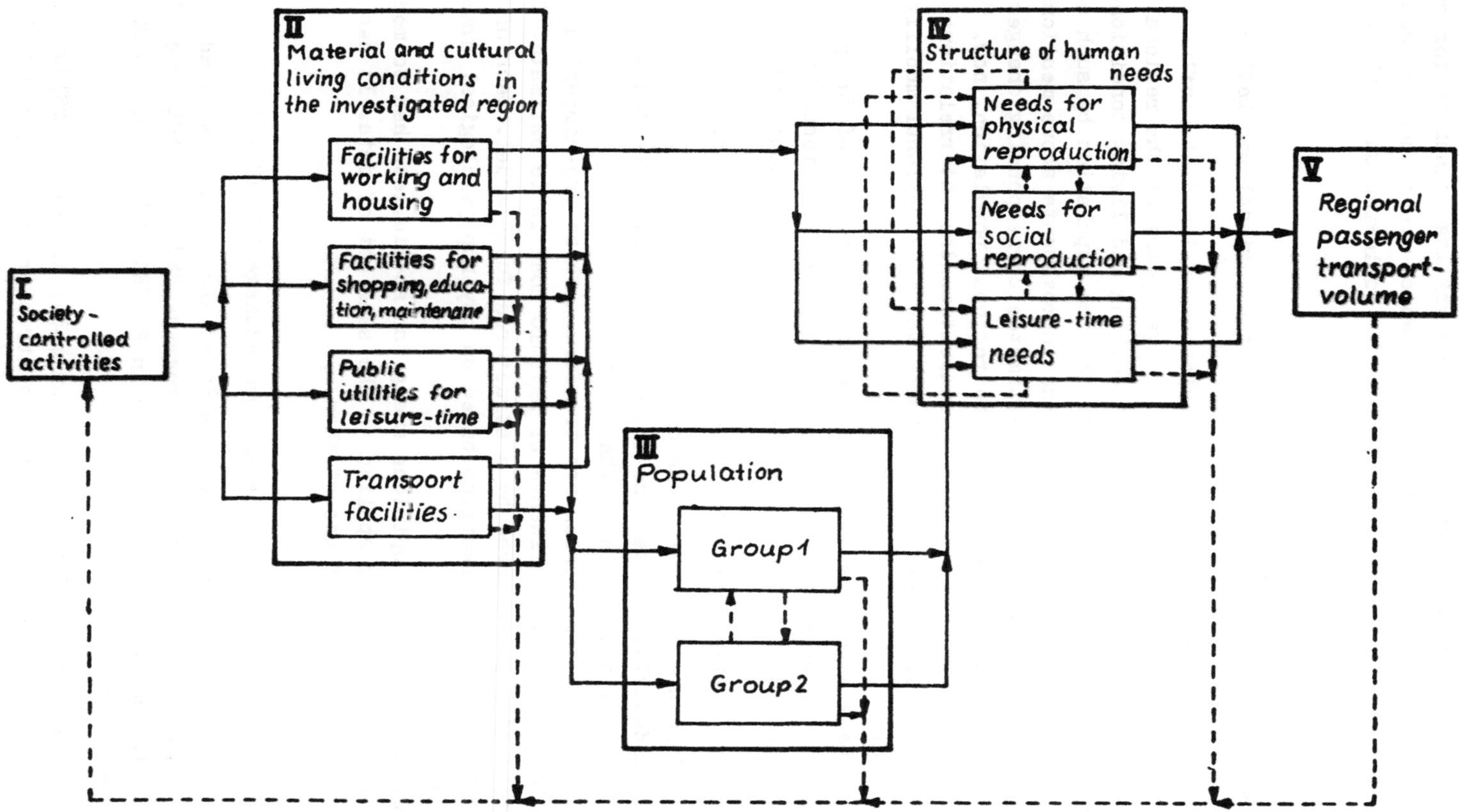

Figure 6: Structure of the fuzzy simulation model for regional planning (schematic depiction)

3.2 Regional planning [4]

A fuzzy simulation model was created for writing scenarios for changes of regional passenger-transport demands which are caused by society-controlled activities. It consists of the following submodels (cf. Figure 6):

- working and living conditions
- demographic facts
- structure of the requirements of the regional population
- mobility

The simulation starts with the calculation of immediate influences for the demographic and socio-structural development of the regional population resulting from improvements of working and living conditions. The next step considers mediate effects for the population by such activities. These effects together with the improvements above-mentioned are the inputs of the submodel which estimates corresponding changes of the structure of the requirements of the regional population, now. The following submodel takes into consideration the interrelations of the requirement itself. With these outputs the interregional mobility is calculated.

Each of these submodels comprimises a defined number of fuzzy variables interconnected by appropriate fuzzy structural relations. The calculation is carried out with discretisized fuzzy time functions.

3.3 Dynamic fuzzy models for psychophysiological diagnosis [8]

These models are used for decision making during the production process in some industrial plants. The diagnostic results have led to some consequences in the schedule of the working day.

Polysymptomatic time series are tranformed to an one-dimensional interpretation space. The influence of psychic load is to estimate. This transformation is achieved by a structured fuzzy model. Each of the submodels is a fuzzy algorithm based on psychophysiological founded vague statements.

Furthermore, different opinions among the specialists had to modelled. These fuzzy models were applied to the same set of data. It has resulted in the preference of one of them.

4. A brief description of the program package FUZSIM [7]

The program package FUZSIM is capable of simulating static and dynamic fuzzy systems. It consists of the programs FSTRUC, FPARAM and FSIM (cf. Figure 7). The programs are written in FORTRAN IV and run on SM4 and PDP-11 minicomputers.

The structure of the fuzzy simulation model is read by FSTRUC. Blocks are used for the description of this structure (cf. Figure 8). The model is regarded as a connexion of such blocks which represent subsystems or elementary systems. Each block has a descriptor according to its task (named by the non-terminal <BLOCK> listed in Table 1) and is characterized by parameters, the output variables, the input and state ones.

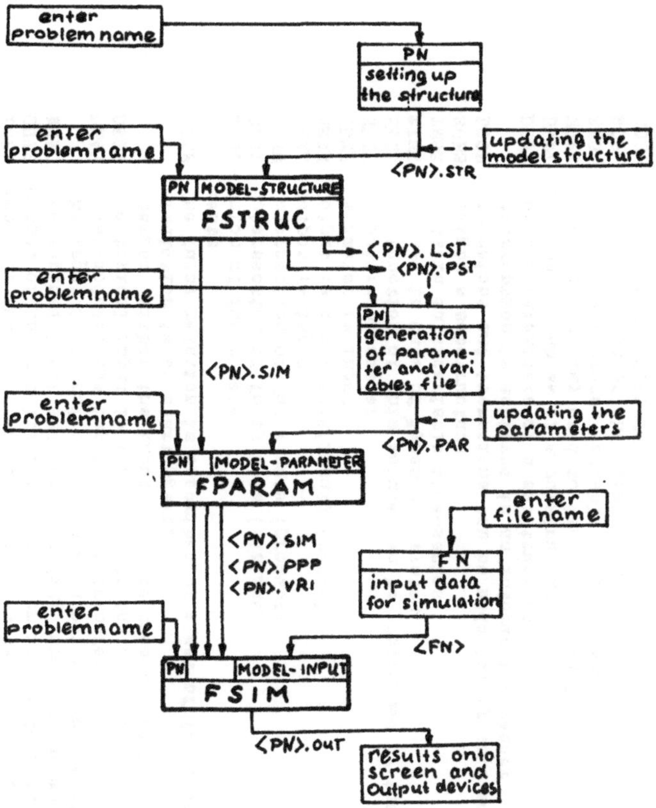

Figure 7 : Schematic representation of **FUZSIM**

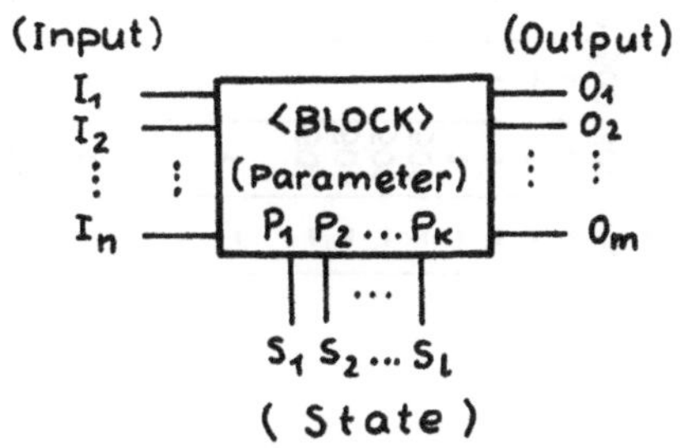

Figure 8: General representation of a block

Table 1: Descriptors of <BLOCK>

symbol		maximal numbers of: parameters	output variables	input variables	state variables
INPT	Input of fuzzy or non-fuzzy data via terminal or other input devices	-	500	-	-
OUTP	Output of the results onto screen or output devices	-	-	500	-
MEMO	The topical input and the preceding input are shown on the screen. The input is transferred to the output and to the state variable	-	1	1	1
DLAY	Delay-module; the input will be transferred to the output at t+1	-	1	1	1
DECN	The topical input is written onto the screen. It has to be decided whether either the input is to be transferred to the output or the simulation is to start again.	-	1	1	-
RELF	The fuzzy relation is described by a set of fuzzy conditional statements	3	1	500	-
DISJ	Union of fuzzy sets	-	1	500	-
CNJI	Intersection of fuzzy sets (minimum)	-	1	500	-
CNJU	Intersection of fuzzy sets (product)	-	1	500	-
EQAL	Output := input	-	1	1	-
NEGN	Output := complement(input) (y=1-x)	-	1	1	-
SETP	Parameter := input	1	-	1	-
SETV	Output := parameter	1	1	-	-
SELF	The function of this module can be programmed by the user	500	500	500	500
TIME	is used in case of dynamical system	-	-	-	-
TMSR	reads in multivariable time-series	1	500	-	-
SUBS	designs a substructure	500	500	500	500
LOOP	Recursions and iterations can be modelled by this module. The interruption is achieved in an interactive manner	-	-	500	-
DEFP	defines parameters of a substructure	500	-	-	-
DEFV	defines variables of a substructure	-	-	500	-
NOTE	16 characters may used to denote a variable				
END	closes a structure, variable or parameter				
EOP	End of the set of the parameters				
EOS	End of MODEL-STRUCTURE				
EOV	End of the set of the variables				

The values of the variables and parameters are fuzzy sets or non-fuzzy values. The parameters remain constantly during a simulation run.
The values of the variable and parameters are read in by FPARAM.
The finite cardinalities of the universes of discourse of the fuzzy sets has to be given. There are four possible types of parameters. Built-in fuzzy sets depicted in Figure 9 acan be used as parameters, too.

$$f_c(x) = 1/(1 + \langle FACTOR \rangle \cdot 0.016 \cdot (c-x)^2)$$

$$f = (f_a \vee \chi_{[a,b]} \vee f_b) - 10^{-4}$$

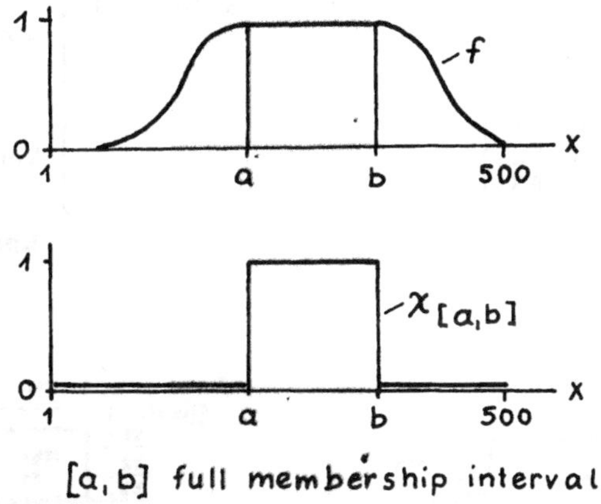

Figure 9 : Built-in fuzzy set

The simulation of the model is carried out by FSIM. The input data can be read via terminal or file. Each included output-block writes the pertinent results onto the screen and into a file.
After finishing a simulation run a new computation can follow by FSIM using the same parameters and the same structure. Using the same structure but new parameters FPARAM has to run before calling FSIM.

An illustrating example is following.

The structure of the model has to be written according to the sytax diagram

MODEL-STRUCTURE

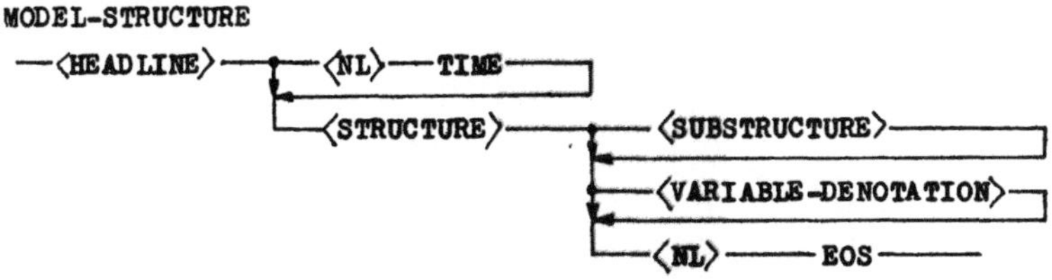

where ⟨STRUCTURE⟩ is defined by

STRUCTURE

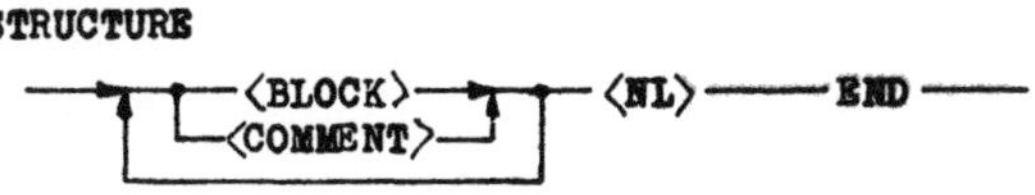

More details are to find in [7] .

Filename: TEST.STR

```
/   ILLUSTRATING EXAMPLE
/   PROBLEM NAME:    TEST                                    HEADLINE
/   DATE: MAY, 1982
//////////////////////////////
INPT (;V1,V2,V3,V4,V5,V6,V7;;)
RELF1(DE,P1,F1;V11;V1,V2;)
SUBS1(DE,P2,P3,F2,F3;V12,V13;V3,V4,V5,V6,V7;)               STRUCTURE
RELF4(DE,P4,F4;V14;V11,V12,V13;)
OUTP (;;V11,V12,V13,V14;)
END
SUBS1(DEF,PA1,PA2,ALG1,ALG2;Y1,Y2;X1,X2,X3,X4,X5;)
RELF1(DEF,PA1,ALG1;Y1;X1,X2,X3;)                             SUBSTRUCTURE
RELF2(DEF,PA2,ALG2;Y2;X4,X5;)
END
NOTE V11 (RESULT OF RELF1 )
NOTE V12 (RESULT OF RELF2 )                                  VARIABLE-DENOTATION
NOTE V13 (RESULT OF RELF3 )
NOTE V14 (RESULT          )
EOS                                                          End of structure
```

Structure

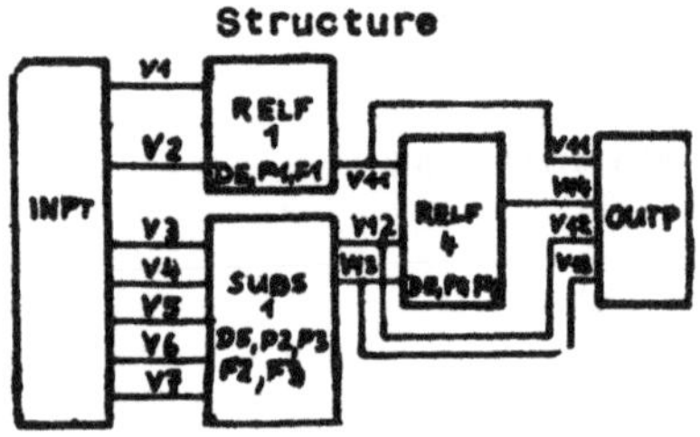

Substructure

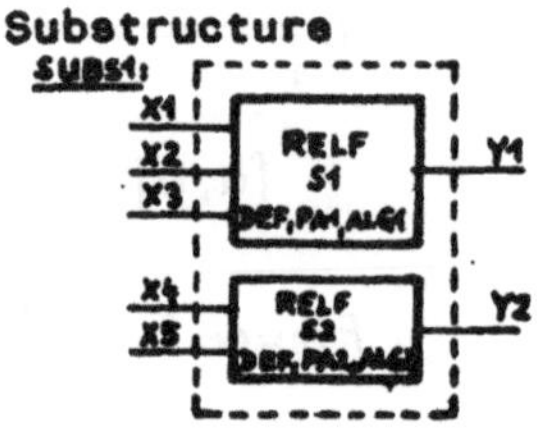

Filename: TEST.PST

```
V1   (**)
V2   (**)
 :
V14  (**)
EOV

DE   (TYP+,**)
END
P1   (TYP+,**)
END
F1   (TYP+,**)
END
 :
P4   (TYP+,**)
END
 :
F4   (TYP+,**)
END
EOP
```

```
  Filename: TEST.PAR
                                            some comments

 V1  (1)
 V2  (1)
 V3  (1)
  :
 V13 (7)---------- number of classes of variable V13
 V14 (5)
 EOV

                          --------this parameter (of  TYP2 ) includes 14
 DE  (TYP2,14)                    of the built-in fuzzy variables
F  A1  425  500  1
F  A2  350  425  1
F  A3  275  350  1
F  A4  225  275  1<---The variable A4 is Fuzzy and its
F  A5  150  225  1    full membership interval is [225,275].
F  A6   75  150  1    The  FACTOR  is equal to 1 (cf. Figure 9).
F  A7    1   75  1
S  B1    1    1  1
S  B2    2    2  1
S  B3    3    3  1<---B3 is a sharp set
S  B4    4    4  1
S  B5    5    5  1
S  B6    6    6  1
S  B7    7    7  1
 END

 P1  (TYP3,2)
   V1  V2  V11 <------Ordering of the variables in accordance
 END                  with fuzzy algorithm F1

 F1  (TYP4,21,3) <----F1 is name of parameter for fuzzy algorithm
     A1,        A1, B1;   with 21 statements and 3 variables
  A1 A2,        A2, B2;<-
  A2 A3,        A1, B2;   \
     :          :   :      \is read IF V1=A1 OR A2 AND V2=A2
  A5 A6,        A7, B7;                       THEN V11=B2
  A7, A1 A2 A3 A4, B7;
     A7,     A5 A6, B7;
 END

 P2  (TYP3,3)
   V3  V4  V5  V12
 END

 F2  (TYP4,46,4)
   A1 A2,        A1,          A1, B1;
      A1,     A1 A2,          A2, B2;
      A2,        A1, A1 A2 A3, B2;
      :          :             :   :
      A5,     A4 A5,          A5, B5;
 END
 :
 EOP

  Filename: INPUT1.

 V1(1)
      382
 END
 V2(1)
      362
 END
 V3(1)
      410
 END          (cont. next page)
```

```
V4(1)
      390
END
V5(1)
      310
END
 :
 :
EOV
```

```
   Filename: TEST.OUT
 /   ILLUSTRATING EXAMPLE
 /   PROBLEM NAME:   TEST
 /   DATE: MAY, 1982
 /////////////////////////////
 V11 : RESULT OF RELF1
CLASSES                                                 interval of
 1 -  7: 1360 9999 8127  762  322  137   75             membership value[0,9999]
 V12 : RESULT OF RELF2                                  with the same
CLASSES                                                 properties of [0,1]
 1 -  5:  451 3378 9999 2808  451
 V13 : RESULT OF RELF3
CLASSES
 1 -  7: 9999 8127 7871  229  229  137   75
 V14 : RESULT
CLASSES
 1 -  5: 1360 9999 8127 2808  322
```

5. References

[1] Elst, G. and B. Straube: Simulation of fuzzy electrical networks. Preprints IFAC Symposium on Fuzzy Information, Knowledge Representation and Decision Analysis, Marseille 1983.

[2] Hansel, N. and B. Straube: Fuzzy modelling for forecasting discharge and water level of large rivers. Optimal Allocation of Water Resources (Proceedings of the Exeter Symposium, July 1982).

[3] Hansel, N. and B. Straube: Application of the method of fuzzy sets to the discharge forecast for large rivers. J. Hydrological Sciences, 7 (1983) 64-76.

[4] Gericke, M. and B. Straube: Changes of reginal passenger-transport demands by society-controlled activities - a fuzzy simulation approach. In: Cybernetics and System Research 2, Trappl (ed.) 1984 pp. 573-579.

[5] Gottwald, S. and. W. Petrycz: Analysis and synthesis of fuzzy controller. Problems of Control and Information Theory, 14 (1984) 33-45.

[6] Peschel, M.: Informationsprozesse und Modellbildung. In: Zur Bedeutung der Information für Individuum und Gesellschaft. Wissenschaftliche Konferenz zum Leibnitz-Tag der AdW der DDR. pp. 135-150. Akademie-Verlag, Berlin 1983.

[7] Straube, B.: FUZSIM - a simulation program package for fuzzy systems. ZKI-Informationen, Berlin, No 2, 1983, 45-63.

[8] Straube, B. and R. Schmitt: Fuzzy algorithms based on vague psychophysiological statements. Preprints IFAC Symposium on Fuzzy Information, Knowledge Representation and Decision Analysis, Marseille 1983.

[9] Workshop on 'the membership function', Brussels 1984.

A PARAMETRIC FUZZY CLASSIFICATION CONCEPT

S.F. Bocklisch; U. Priber [1)]

1. CHARACTERISTIC OF SYSTEMS

Linear and nonlinear systems can be characterized in a state space, may be a phase plane, by state trajectories like focus, center, node, saddle or different types of limit cycles (see fig. 1).

If we investigate a complex and complicated system (a global system), we do not know an exact description by differential equations and also not the right state variables. We have only input and output informations and more or less expert knowledge about the system. In such cases we should use all informations about the system we can catch as indirectly state variables. That are primary informations from measurements and expert consultation and secondary informations that are results of signal processing (Fast Fourier Analysis, regression analysis, parameter estimation and so on). All informations together we call features and the state space is now a feature space.

Now for a large scale system with nonlinear dynamic subsystems we can find typical system situations in that feature space. Such systems have some different stable points. Fig. 2a-d shows types in the time domain and in the two dimensional feature space:

- one stable point
- a cyclic or rhythmic behaviour we know from biological systems with the two states or classes K1 characterized by the expression "m_1 is low and m_2 is low" and K2 characterized by "m_1 is high and m_2 is high"
- two stable system situations separated by the separatrix S (an instance is a typical failure or disease illustrated by class K3)
- an evolutionary system with a nonreversible movement that means a good or a bad trend

In an extremely way catastrophic systems show the phenomen of different stable situations. Such systems have no steady state between different stable points.

We believe that classification is a suitable instrumentarium for modelling such systems.

---

1) Academy of Sciences of the G.D.R., Institute for Mechanics, DDR-1199 Berlin, Rudower Chaussee 5

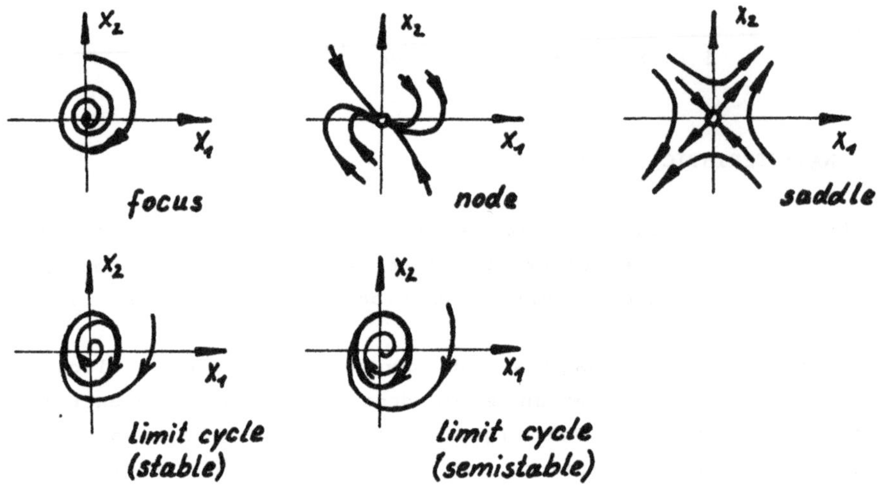

Fig. 1. Types of phase portraits

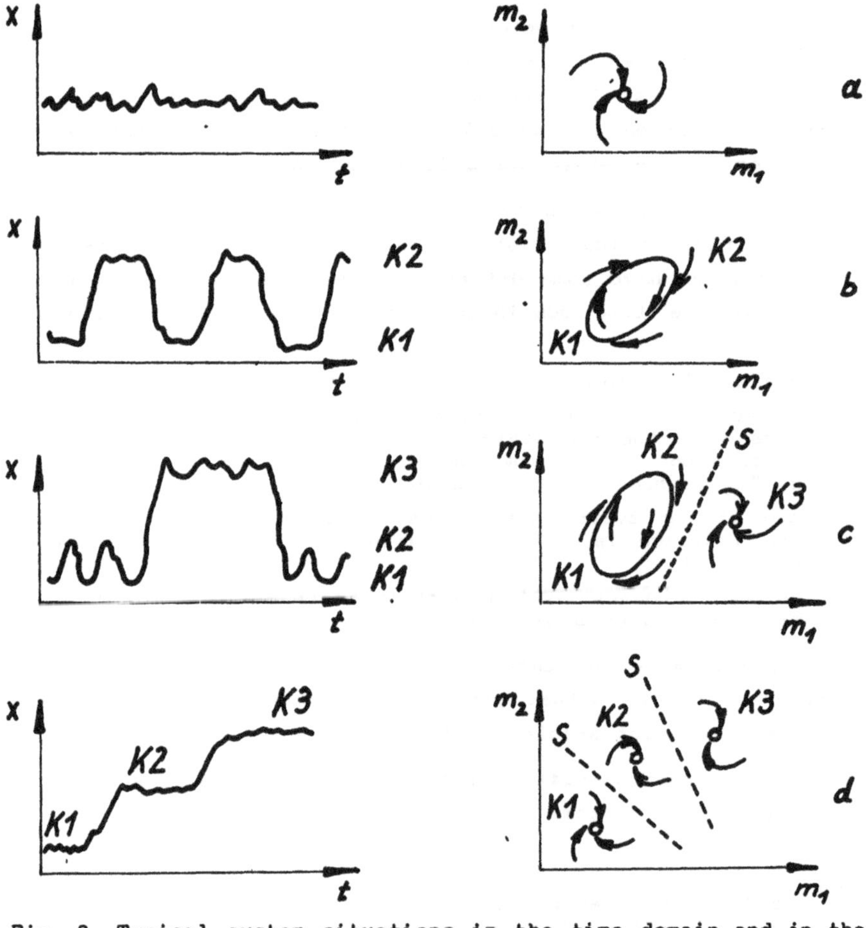

Fig. 2. Typical system situations in the time domain and in the feature space

2. CLASSIFICATION

Classification is a subdivision of a set of elements or objects specified by features into a set of different subsets each with special importance. Objects can be animated or lifeless concreta like persons, animals, also industrial plants, buildings, machine devices, or abstracta like philosophies, opinions, methods, states or solutions.

Features are direct or indirect measurements of quality and quantity or derived values for instance parameters. They are metric or nonmetric ones.

Base of every classification is a set of objects

$$X_{\bullet} = (X_1, X_2, \ldots, X_I) \qquad (1)$$

and special for vector classification every object X_i is described by a feature vector

$$X_i: \quad \underline{m}_i = (m_1, m_2, \ldots, m_N)_i \qquad (2)$$

The possibilities of classification are that of fundamental ordering and structure buildung. It is a model buildung method of experimental process analysis.

<u>Structure building</u>

For structure building we prefer pattern methods. Advantages we see in the following facts:

a) The factor of data reduction is very high. Together with a parametric concept for class description it is possible to construct very effective software instruments.

b) Fuzziness is controlable (by operator or automatically) by dialoque or within the algorithms. So we can choose a special level of fuzziness.

c) By using ordinal scales an extrapolation to interclass regions is possible.

d) Every class membership function is adapted to the data base. So we get more information on fuzziness additional to the expert.

e) In the structure building step we get a cluster configuration. Cluster found by cluster methods are of natural type. It is desirable that natural classes agree with semantic ones that means they are of practical importance. But often they are not identically. Then we get in general a larger number of natural than semantic classes. That increase of specification provides an increase of classifier quality. Furthermore the natural substructure can use as a hypothesis for the expert.

We know several types of cluster methods. They all have special advantages and disadvantages. For instance it is very difficult to find the two classes of configuration 5 in fig. 3 with a hierarchical cluster method. On the other hand it is difficult to find the two classes of configuration 2 with a nonhierarchical (sequential or iterative) method, because the criterion is that of minimal variance.

We notice:

- Cluster methods are objective structure building procedures.
- A second possibility is to define a structure by an expert, a teacher, as subjective structure building.

With every structure we define also a level of accurancy or uncertainty. It is very important to determine the level of view. So it is possible to get global or detailled reflections.

Every class itself may be interpreted as a submodel of the general problem.

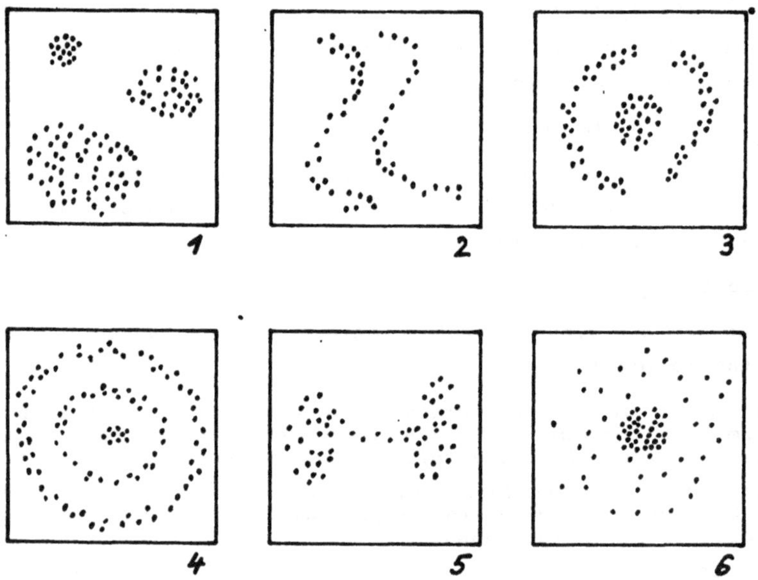

Fig. 3. Examples of cluster configurations

Description of the cluster structure

Classical nonfuzzy methods for description of nonoverlapping disjunct clusters are for instance

a) the enumeration of all objects
b) informations on the boundary lines
c) class description by hyperquaders or more flexible functions more or less adapted
d) declaration of intervals
e) linear or nonlinear separation planes.

A disadvantage of some kinds is that such a classifier yields only crisp membership values (0 or 1) and also a well defined result for an object with a great distance from every class. That is very risky because there could be an unknown class.

So we use a fuzzy concept that is suitable for description overlapping classes and is able to refuse objects. That is possible because the membership to every class in such a case is very small.

3. CONCEPT OF AN UNIFORM PARAMETRIC MEMBERSHIP FUNCTION

During the last years we developed an uniform parametric membership function conception $\mu(\underline{m};\underline{p})$ (see /4/).

We prefer it because

- Todays available computers work nonfuzzy. Therefore they can not manipulate membership functions in a fuzzy way. Parameter level is better for an effective handling, however special aggregation algorithms are needed.
- We get an extremely high data reduction factor with the parametric concept but it is necessary to use a special model. So the concept is very suitable for description of very high dimensional membership functions (practical to 100).
- The decomposition of separate informations to parameters is possible (such informations are position or fuzziness).
- The function of the same type can be used for describing both elementary informations (objects) and global ones (classes), $\mu_e(.)$ repectively $\mu_g(.)$.

Our concept is a generalization of Aizermans potential function. Similar types are also used to describe niches in ecology.

The aggregation operation for the elementary membership functions is principally a disjunction one. It is done on the parametric level. Heuristically it is used because of every $\mu_e(.)$ is looked as a rough estimation of a complete cluster.

The membership function has in normally position (origin, without rotation) in a n-dimensional feature space the form

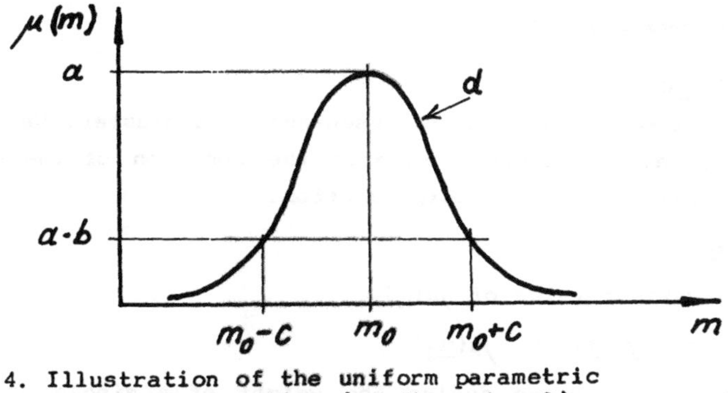

Fig. 4. Illustration of the uniform parametric membership function (onedimensional)

__Parameter c__

c is the boundary of the class, that is the position of the extremal objects in projection to the gravity axis. So c represents the greatest distance of an observed object from m_0.

$$c > 0 \tag{8}$$

__Parameter d__

d is an information parameter for decrease in membership with increasing distance from m_0. It is determined from object distribution. It yields

$$d > 0 \tag{9}$$

and for $d \longrightarrow \infty$ we get the ordinary nonfuzzy membership function as a special case.

All these parameters (and the angles for adaption in the space) are determined for every class. Especially the parameters b, c and d may be different on left and right side of the function (labled by the indizees l and r).
So every class is described by a number of 7*N parameters (N the total number of features).

$$\mu(\underline{m};a,\underline{b},\underline{c},\underline{d}) = a \left[1 + \sum_{n=1}^{N} (1/b_n - 1) \mid m_n/c_n \mid^{d_n}\right]^{-1} \tag{3}$$

Generally for the description of a class it is moved into gravity centre $\underline{m}_0$ and is rotated by $\underline{\varphi}$. So we get a set of determinant parameters for every class.

Parameter designation is realized by projection of the elementary membership functions describing all objects of a class into the plane, streched by gravity axis (determined before) and ordinate. By means of that operation an independence is reached from other dimensions. So we get an efficient description in a feature space theoretically with an unlimited number of dimensions. For parameter discussion we use an onedimensional function (see fig. 4):

__Parameter m_0__

m_0 is the position of the representant of a cluster. We use the gravity point and there lies also the position of the maximum value of the class membership function.

__Parameter a__

a is the maximum value of $\mu(\underline{m})$.

$$a = \max \mu(\underline{m}) = \mu(\underline{m}_0) \tag{4}$$

As a normalization we define the weight of a single object to a = 1.

Determination of a is

- done by an expert according to the importance of the class (for instance a dangerous situation in a plant or a special disease) or
- is grounded on statistical experimental results, in that case a is a function of the number of objects I_k in the k-th class

$$a = a(I_k) \tag{5}$$

a has the special importance for valuation of a class. So it is used for updating timevariant classifiers by means of learning procedures (see /6/).

Parameter b

b is a factor that determines the value of membership function at the sharp boundary c of the class

$$\mu(m_0 \pm c) = a*b \tag{6}$$

$$0 < b \leq 1 \tag{7}$$

For the complete global membership function we get

$$\mu_g(\underline{u};a,\underline{b},\underline{c},\underline{d}) = a\cdot\left\{1 + 0.5\sum_{n=1}^{N}\left[(1-\mathrm{sgn}(u_n))\cdot(1/b_{ln}-1)|u_n/c_{ln}|^{d_{ln}} + (1+\mathrm{sgn}(u_n))\cdot(1/b_{rn}-1)|u_n/c_{rn}|^{d_{rn}}\right]\right\}^{-1} \tag{10}$$

with

$$\underline{u} = A * (\underline{m}) \quad \text{(A: affine transformation for translatoric and rotatoric adaption)} \tag{11}$$

and signum-function

$$\mathrm{sgn}(u) = \begin{cases} -1 & \text{for } u < 0 \\ 0.5 & \text{for } u = 0 \\ +1 & \text{for } u > 0 \end{cases} \tag{12}$$

The elementary membership function describes a single object and has the symmetrical form:

$$\mu_e(\underline{m};\underline{m}_i,c_e) = \left[1 + \sum_{n=1}^{N}((m_n - m_{n,i})/c_e)^2\right]^{-1} \tag{13}$$

with

$\underline{m}_i$ position of the object

c_e parameter for elementary fuzziness of the object

This function is derived from $\mu_g(.)$ by the special definitions:

$\underline{m}_0 = \underline{m}_i$ $\quad a = a_e = 1$

$b_l = b_r = 0.5$ $\quad c_l = c_r = c_e$ $\quad d_l = d_r = 2$

4. LEARNING AND WORKING PHASE

The block scheme of fig. 5 gives an overview of the classifier-building and using strategy.

Structure building and description we call the <u>learning phase</u>. Result is a timeinvariant classifier, that is a algorithm or a device.

In the <u>working phase</u> the classifier is used off line or on line. So for a given object a sympathetic vector is determined. With its components it gives a (different) membership value (value of affinity) to <u>every</u> class. We destinguish

- the absolute sympathetic vector $\underline{\mu}(m_i)$
- the relatively sympathetic vector $\underline{\mu}'$, that is the absolute, devided by the class maximum membership a_k. So it has values between 0 and 1

$$\mu'_k = \mu_k / a_k \tag{14}$$

- the elementary sympathetic vector $\underline{\mu}''$, that has a sum of its components of 1 (or 100 percent). It has the meaning, that a single object gives parts of its general energy to every class.

$$\mu''_k = \mu_k \Big/ \sum_{k=1}^{K} \mu_k \tag{15}$$

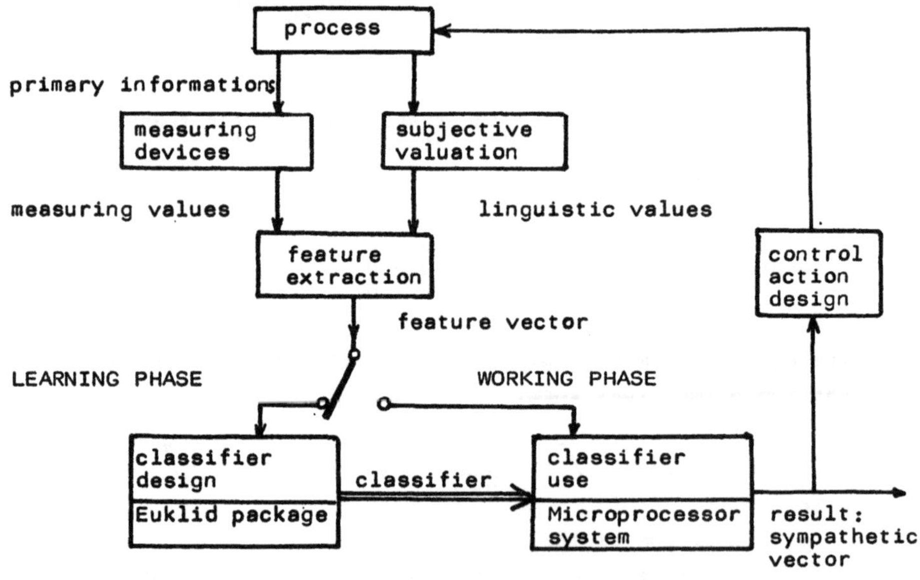

Fig. 5. The strategy of classifier building and using

Normal the static identification is used. But of widespead interest are dynamic considerations. Then the sympathetic vector becomes timevariant. The great practical importance is to look for trends, the prognostic development of the system and the determination of evolutionary laws.

In case that we have observations from time variant problems, we need a self learning algorithm. It should be able to update all parameters of the classes, to generate new classes, to fuse or to split ones.

If we summarize there are four typical ways for using a classifier:

a) static identification with a timeinvariant classifier
b) dynamic identificatin with a such one
c) static identification with a timevariant and
d) dynamic identification with a timevariant classifier.

In the last case the time horizon of changes within the system must be smaller as that of growth in knowledge about the general problem that provides the movement of class structure.

6. SOFTWARE INSTRUMENTS

The available computer software is devided into two parts that means two levels.

The computer software package EUKLID, a complex working and simulation instrument

It enables the use of all procedures and algorithms mentioned above for modelling of real systems, for construction of fuzzy classifiers and for use of practical problems. Some of its characteristics are

- the dialogue parts and the problem relevant control language
- the modular and expansible structure based on FORTRAN-subroutines
- a workable problem size of about 1000 learning objects each characterized by 100 features per block.

The efficiancy of the system is proved in a wide range of application, especially by flexible combination of its moduls.

Some of the most important moduls of the package are:

- the control modul including message generation and dialogue control
- data handling inclusive normalization ect.
- clustering methods
- fuzzy description of classes by membership functions, computing of a fuzzy classifier also for use in a microprocessor system
- feature reduction and structure valuation
- fuzzy identification, computing of the sympathetic vector
- learning classification and classifier updating
- multivariate time series classification and modelling
- ect.

Microcomputer software for process near fuzzy identification on-line or off-line

This software includes BASIC-algorithms, allocated in the working step of the classification system, for instance for identification and classifier updating. The microprocessorsystem or personalcomputer is fed with the complete fuzzy classifier calculated by EUKLID-package. After that it works independent.

Grounded on a real time processing continous identification allows to determine trends and to realize feedbacks to the practical process (process control, medical therapy, planing of repairs, management decisions and so on).

APPLICATIONS

Fuzzy classification is applicable in a wide range of process identification, modelling, decision making and so on in very different technical and nontechnical application areas. Some special examples we treated were (see /1/-/7/):

- quality inspection, special of light emission diodes
- description of intensity distributions, for instance regional planing of different services in regions, description of pollution, also as a function of time
- recognition of standardized objects (like work pieces) for robot control by means of extremely low resolving sensors
- description of aging effects of cristal filters for identification unknown ones and estimation of trends
- classification of animal blood pressure curves under stress situations
- diagnosis of the human arteriel blood circuit system
- machine tool and wear fault diagnosis
- process inspection in boring and milling machines
- adaptive control with a fuzzy classifier controller concept

REFERENCES

/1/ Fortschrittsberichte der GDR Working Group for Fuzzy Sets and Systems, Akademie der Wissenschaften der DDR, FB Mathematik und Informatik, Band I bis VI (1977-1983)

/2/ Bilz,F., Bocklisch,S.F., Meyer,W., Peschel,M., Straube,B.: Fuzzy concepts for systems identification of biomedical systems, Proc. IFAC-Congress, Helsinki (1978) 499-507

/3/ Peschel,M., Bocklisch,S.F., Meyer,W., Straube,B., Richardt: Daten- und Verhaltensanalyse mit Klassifikationsmodellen, Österreichische Studiengesellschaft für Kybernetik, Rep.23 (1981)

/4/ Bocklisch,S.F.: Experimentelle Prozeßanalyse mit unscharfer Klassifikation, Diss.B, TH Karl-Marx-Stadt (1982)

/5/ Bocklisch,S.F., Lange,E., Mann,S., Priber,U.: Ein unscharfes Klassifikationssystem, Wiss. Schriftenreihe der TH Karl-Marx-Stadt (1982) H.13, 19-36

/6/ Mann,S.: Ein Lernverfahren zur Modellierung zeitvarianter Systeme mittels unscharfer Klassifikation, Diss.A, TH Karl-Marx-Stadt (1983)

/7/ Programmsystem zur unscharfen Klassifikation, Nutzerhandbuch, Akademie der Wissenschaften der DDR, Institut für Mechanik, Karl-Marx-Stadt (1984)

FUZZY INFERENCES FOR THE ANALYSIS OF QUALITATIVE BEHAVIOUR

Peschel M., Straube B.[1)] and Mende W.[2)]

1. Introduction

Usually the qualitative behaviour of highly aggregated systems is studied on the base of a description by a set of ordinary differential equations. As alternative with some advantages we propose here to apply the method of fuzzy inferences (fuzzy if-then-statements). Because the main approaches for expert systems use if-then-statements we get then the possibility to use all kinds of expert knowledge to support macro-modelling.

The method proposed here must be organised in three stages. At first we need a description of the macrosystem by a socalled causal graph, which describes in a very rough form the if-then-connections between all pairs of variables.

Then we implement the causal graph relations with the help of membership-functions in the sense of the theory of fuzzy sets and systems. By this method in fact the qualitative behaviour in terms of trajectories of time-functions is reduced on the qualitative behaviour of class transitions.

For analysing the qualitative behaviour we apply any fuzzy simulation method, in our case FUZSIM elaborated by B. Straube.

At the end of the paper the validation of our approach is demonstrated by some examples where we applied this approach on some simple systems described by ordinary differential equations.

2. Causal graphs for the description of macromodels

A causal graph is a rough description of a macromodel of a system. The nodes of the graph are the essential variables which are taken into account, where as the arches between two nodes respective the "autocatalytic" nodes mark existing interactions between the corresponding variables. The causal graph contains only orientated arches meaning the interaction is always accounted in the form of cause and response.

---

1) Academy of Sciences GDR, Scientific Department Mathematics/ Informatics

2) Academy of Sciences GDR, Scientific Department Geo- and Cosmos Sciences

Fig. 1 shows an example of a causal graph.

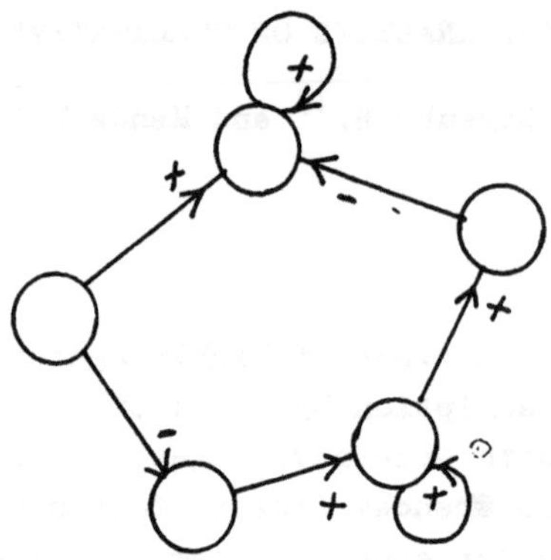

Figure 1

Usually the inferences within a causal graph are according to "proportional thinking" reflected by monotonous resp. antitonous operators working on the variable increments in the following sense:

+- inference of y on x : sig (Δ y Δ x) > 0 monotonous

-- inference of y on x : sig (Δ y Δ x) < 0 antitonous

The character of the inference + or - is notated at the arches of the causal graph.
A causal graph given in such a form can be immediately implemented using for example a block-orientated simulation language as for example DYSYS (Dynamical System Simulation), elaborated by H. Bossel /1/.

Fig. 2 shows a concrete example for a causal graph, which was taken

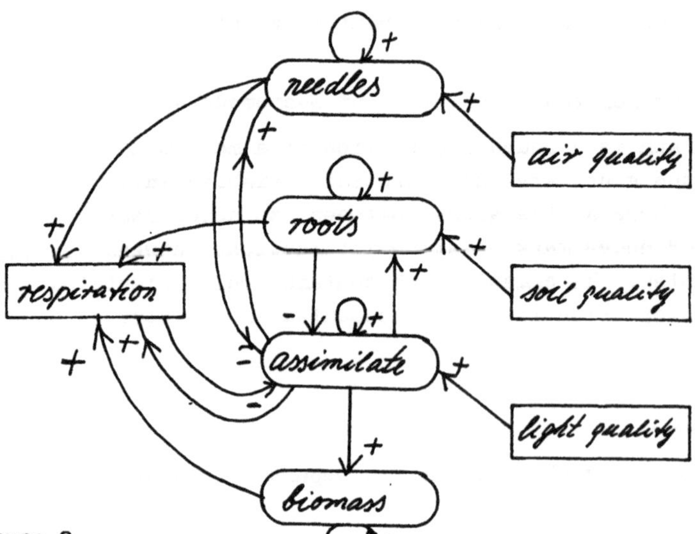

Figure 2

from ecology /1/, the macromodel of a tree (fir-tree).
In the concrete simulation based on such a causal graph some refinements are used, for example

- According to the interation tact (month for example) a time-variable input of light and temperature is applied
- needle-age classes 1...8 are introduced
- liveliness of the tree according to its age will be evaluated and taken into account by some model parameters.

3. Fuzzy implementation of a causal graph

The traditional implementation of a causal inference in a given reference-point (y_o, x_o) is determined by "proportional thinking" in the following way

$$\Delta x_o = K (y_o, x_o) \; \Delta y_o$$

If $K > 0$ we meet a +-inference, if $K < 0$ we meet a - -inference.
If we possess such a local linear model, it is quite natural to get a fuzzy implementation of the causal inference.
In such a case it would be necessary to construct a local membership-function $\varphi(y_o, x_o, \Delta y_o, \Delta x_o)$ supported in its geometrical form by the local linear model.

A more consequent way for fuzzy implementation which does not need apriori a local linear model is based on the fuzzy quantification of the input and output variable, for example in the following way

y_o : small, medium, large
x_o : small, medium, large

In this case the fuzzy inference knowledge can be asked at least partially from an expert.
Similarly also an autocatalytic connection, meaning an influence of a variable x on itself, can be implemented by the fuzzy approach by a membership-function of the form

$$\varphi(x_o, \Delta x_o)$$

which determines by a suitable decision rule to the reference value x_o of the variable x a reasonable increment ΔX_o in the next step.
The simulation procedure on the base of the fuzzy implementation just described can be performed with different approaches.
We propose some of the possible fuzzy simulation methods. The main idea for the simulation is the vector-iteration parallely for all nodes of the causal graph

- recursive decision for all Δx_i using the local membership-functions above introduced making use of the input increments.

- shifting of the reference points according to the computed increments for the outputs of the fuzzy inferences

$$x_i \rightarrow x_i + \Delta x_i$$

But, as can be seen from the causal graph in Fig. 1 there exist usually nodes with some imputs.
The problem arises, how we should compute the corresponding output increment on the base of some binary inferences.
There arise in the simulation task some possibilities.
One possibility is to take into account the binary inferences of a given node one after the other, gather the corresponding increments of the output variable and add them may be taking into account weights which are given by the used binary fuzzy inferences (membership-values).
Only after having accumulated in such a way the single increments into an aggregated increment Δx_i we perform the shift of the reference value x_i.
On the other hand we could in a subroutine perform a shift already for each of the binary inferences producing thus a large shift after having run through all the input variables of the considered node in a predetermined sequence.
We could also avoid vector iteration of the variables of the whole causal graph if we fix a certain path meeting in one cycle only once every node of the graph and producing the corresponding fuzzy increments for every node of the graph according to the predetermined path through the nodes of the graph individually.

4. The dominating role of Lotka-Volterra equations for Macromodels

It was shown in /2/ that all macromodels of aggregated systems which are described by a model in the form of a set of ordinary differential equations can be transformed in such a way that we get a standartisized description by the Lotka-Volterra equations /3/

$$F x_i = \sum_j G_{ij} x_j + \sum_S H_{is} y_s$$

with $F = d\ ln/dt$ the logarithmic derivative.
By such a description a corresponding causal graph is uniquely determined. Obviously we get the result

$G_{ij} = 0$ no inference

$G_{ij} > 0$ +- inference

$G_{ij} < 0$ -- inference

Similarly the binary inferences between x_i and y_s are given by considering the matrix H_{is} instead of G_{ij}.

In 5 we demonstrate the fuzzy implementation method for systems modelled by Lotka-Volterra equations.

5. Fuzzy approach to Lotka-Volterra systems

In the following one possibility for implementation of autonomous Lotka-Volterra equations

$$\dot{x}_i / x_i = \sum_j G_{ij} x_j \qquad i=1(1)n \quad j=1(1)n+1$$

$$x_{n+1} = \text{const (resource)}$$

by the fuzzy approach is demonstrated.
We perform the following operations of discretization:
Time t : $x_i(k+1) = x_i(k) + \Delta x_i(k)$

$$x_i(k) + h\,\dot{x}_i(k) = x_i(k)\,(1+h\,\dot{x}_i(k)/x_i(k))$$

The variables x are substituted by fuzzy sets as follows:

ordered classes A, B, ... ,G, H
fuzzy sets "very small" .... "very large"

G_{ij}: Instead of the values G_{ij} fuzzy parameters are introduced by their elementary membership-functions according Fig. 3

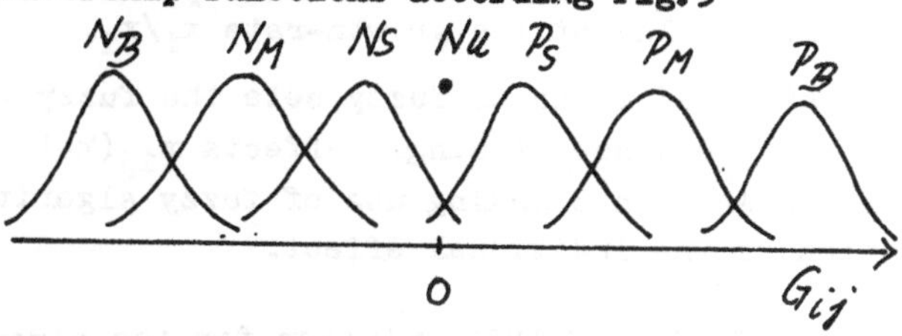

Figure 3

Nu is a δ-singleton: noninteraction between x_j and x_i ($G_{ij}=0$) is completely preserved.

We introduce the partial relative growth rates

$$\dot{x}_{ij} / x_i = G_{ij} x_j \text{ and get then } \quad \dot{x}_i/x_i = \sum_j \dot{x}_{ij}/x_i$$

$\dot{x}_i/x_i$,: We discretisize the rates also by elementary membership-functions
$\dot{x}_{ij}/x_i$ according to Fig. 4

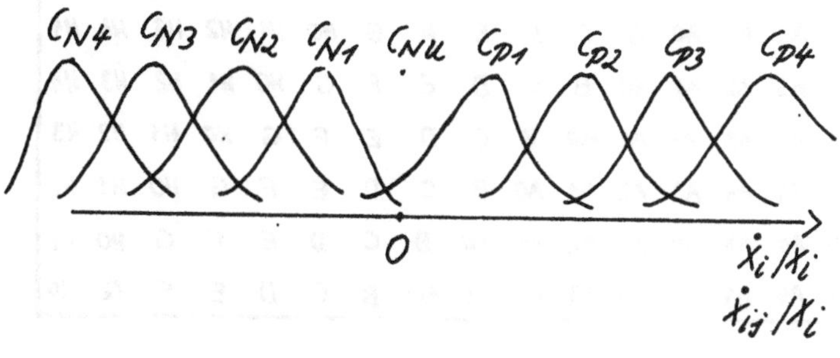

Figure 4

All computations necessary for the simulation of the fuzzified descriptions are performed with the help of two fuzzy algorithms GX and RATE.
The algorithm GX is given by the fuzzy relation in Table 1

Table 1: Operational table for GX

GX	A	B	C	D	E	F	G	H
PB	CP1	CP1	CP2	CP2	CP3	CP3	CP4	CP4
PM	CP1	CP1	CP1	CP1	CP2	CP2	CP3	CP3
PS	CP1	CP1	CP1	CP1	CP1	CP1	CP2	CP2
NU	CNU	CNU	CNU	CNU	CNU	CNU	CNU	CNU
NS	CN1	CN1	CN1	CN1	CN1	CN1	CN2	CN2
NM	CN1	CN1	CN1	CN1	CN2	CN2	CN3	CN3
NB	CN1	CN1	CN2	CN2	CN3	CN3	CN4	CN4

GX produces for given classes of x and G the fuzzy output G x.
The partial relative growth rates $\dot{x}_{ij}/x_i$ after their fuzzy quantification according to Fig. 4 must be combined with each other to get the fuzzy value of the growth-rate $\dot{x}_i/x_i$.

Instead of addition of fuzzy sets the fuzzy value $x_i(k+1)$ is calculated by a sequence of single effects $x_{ij}(k+1)$ as pointed out in general form in section 3 making use of fuzzy algorith RATE for fuzzy-accumulation of the individual effect.

Table 2 shows the fuzzyrelation for the algorithm RATE

Table 2: Operational table for fuzzy algorithm RATE

RATE	A4	A3	A2	A1	A0	B	C	D	E	F	G	H0	H1	H2	H3	H4
CP4	A0	B	C	D	E	F	G	H0	H1	H2	H3	H4	H4	H4	H4	H4
CP3	A1	A0	B	C	D	E	F	G	H0	H1	H2	H3	H4	H4	H4	H4
CP2	A2	A1	A0	B	C	D	E	F	G	H0	H1	H2	H3	H4	H4	H4
CP1	A3	A2	A1	A0	B	C	D	E	F	G	H0	H1	H2	H3	H4	H4
CNU	A4	A3	A2	A1	A0	B	C	D	E	F	G	H0	H1	H2	H3	H4
CN1	A4	A4	A3	A2	A1	A0	B	C	D	E	F	G	H0	H1	H2	H3
CN2	A4	A4	A4	A3	A2	A1	A0	B	C	D	E	F	G	H0	H1	H2
CN3	A4	A4	A4	A4	A3	A2	A1	A0	B	C	D	E	F	G	H0	H1
CN4	A4	A4	A4	A4	A4	A3	A2	A1	A0	B	C	D	E	F	G	H0

The rating up of the individual increments at every node of the graph is done with the help of the following calculation scheme:

step 0 : row i; put $j = 0$
1 : put $x_{io}(k+1) = x_i(k)$
2 : $j = j + 1$
3 : $(x_j(k), G_{ij}) \xrightarrow{GX} (\dot{x}_{ij}/x_i)(k))$
4 : $((\dot{x}_{ij}/x_i)(k)), \dot{x}_{i,j-1}(k+1)) \xrightarrow{RATE} \dot{x}_{ij}(k+1)$
5 : if $j < n+1$ then go to step 2 else
6 : put $x_i(k+1) = \dot{x}_{ij}(k+1)$
7 : all rows?

On page 163 are given five examples of low-order Lotka-Volterra equations with their simulation by the fuzzy approach just described. As can be seen by deterministic simulations the received results from the fuzzy approach very well reflect the qualitative behaviour of the considered systems.

References

[1] Bossel, H.: Umweltdynamik, Te-Wi-Verlag, München 1985

[2] Peschel, M. and Mende, W.: Do we live in a Volterra World? Akademie-Verlag Berlin, Springer-Verlag New York - Wien - Heidelberg 1985

[3] Volterra, V.: Lecons sur la theorie mathematique de la lutte pour la vie, Gauthier-Villars, Paris 1931

[4] Peschel, M.: Modellbildung für Signale und Systeme. VEB Verlag Technik, Berlin 1978

Example 1: $\frac{\dot{x}_i}{x_i} = \sum_j^n G_{ij} x_j$, $i = 1(1)n$, $j = 1(1)n+1$, $x_{n+1} = \text{const.} = 1$

$$G = \begin{pmatrix} -0.4 & 1.0 & -0.6 & -1.0 & 1.0 \\ -1.0 & -0.4 & 1.0 & -0.6 & 1.0 \\ -0.6 & -1.0 & -0.4 & 1.0 & 1.0 \\ 1.0 & -0.6 & -1.0 & -0.4 & 1.0 \end{pmatrix}$$

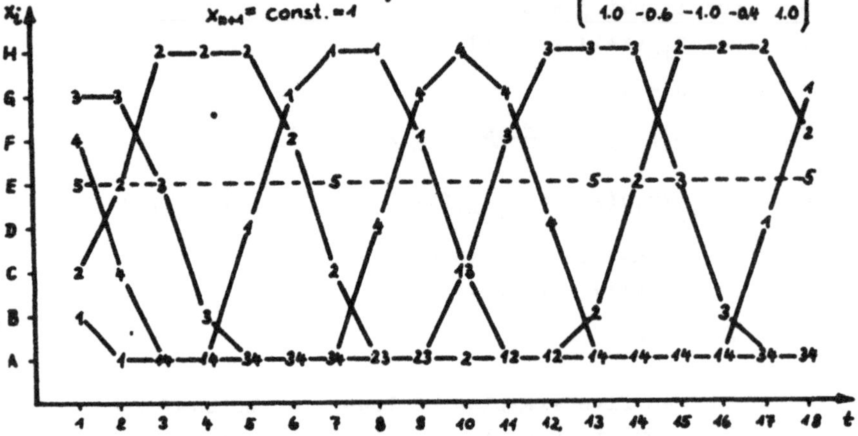

Ex.2: $G = \begin{pmatrix} -1 & -1 & 2 & 1 & 0 \\ 2 & 1 & 1 & 0 & 1 \\ 1 & 0 & 2 & 1 & 1 \\ 3 & 2 & 0 & -1 & 1 \end{pmatrix}$

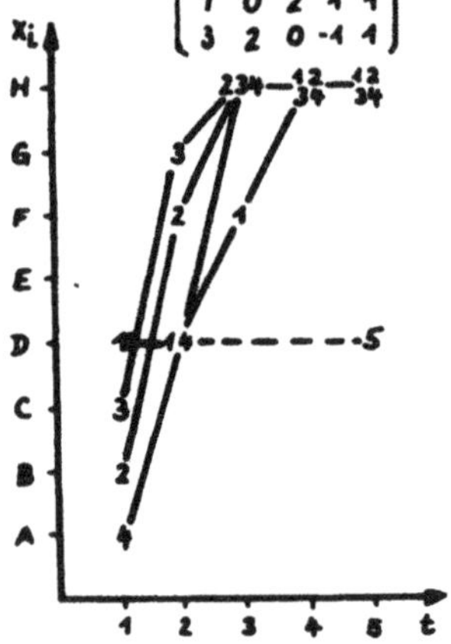

Ex.3: $G = \begin{pmatrix} 1 & 1 & -2 & -1 & 0 \\ -2 & -1 & -1 & 0 & -1 \\ -1 & 0 & -2 & -1 & -1 \\ -3 & -2 & 0 & 1 & -1 \end{pmatrix}$

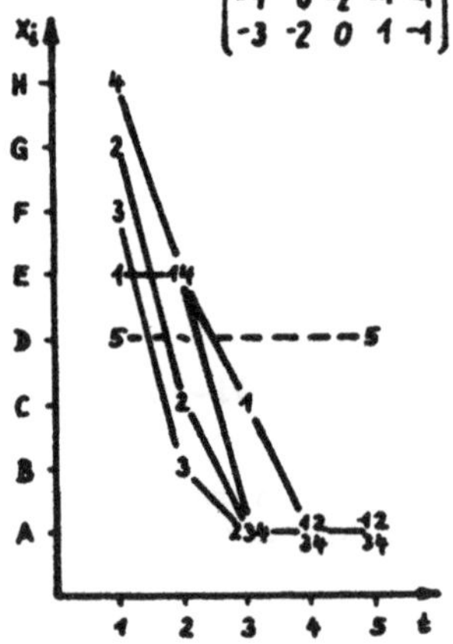

Ex.4: $G = \begin{pmatrix} -3 & 1 & 1 & 0 & 1 \\ 1 & 0 & -3 & 1 & 1 \\ 4 & -1 & -4 & 1 & 0 \\ -5 & 2 & 5 & -1 & -1 \end{pmatrix}$

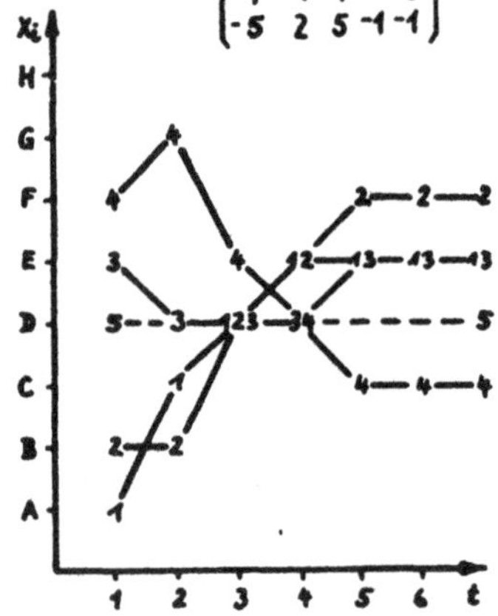

Example 5: $G = \begin{pmatrix} -K & K & 0 & 0 & 0 \\ -K & 0 & K & 0 & 0 \\ -K & 0 & 0 & K & 0 \\ -K & 0 & 0 & 0 & K \end{pmatrix}$

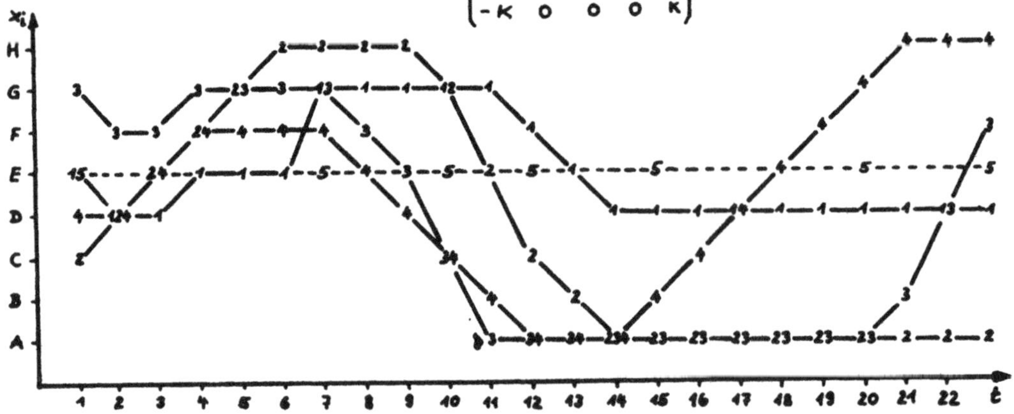

SOME REMARKS TO FUZZY CLUSTER ANALYSIS

Bernd Straube [1])

Abstract. It is pointed out that a careful model building with respect to cluster analysis has to be carried out. Some of the problems encountering are discussed. A summery of the author's applications is presented.

1. Introduction

Cluster analysis is a tool for recovering structures in a given set of objects under the consideration of their internal relationships. Moreover, it is possible to imply a structure within the real system the data are taken from.

The set of n objects $P = \bigcup_{i=1}^{n} \{P_i\}$ is given. Furthermore, the relationships between the the objects are also presupposed. They are expressed by similarity values $s(P_i, P_j)$ derived from a similarity measure σ which is defined on the set $P \times P$. These similarity values have to be known explicitly (e.g. by means of a matrix: 'distance matrix') or implicitely (e.g. by means of an analytical formula or an algorithm). In the case of implicitely given similarity relations metrics are mostly employed. Note that there is an opposite behaviour between similarity and distance.

The structure is found when the set P is decomposed into subsets called clusters.

The decomposition is to be achieved according to the following criteria:

- Objects belonging to the same cluster should be as similar as possible.
- Objects belonging to different clusters should be as dissimilar as possible.

These criteria lead to a task of polyoptimization.
In principal, the known cluster techniques differ one another in solving this task. In this respect, heuristic points of view play a large part.

By means of cluster analysis every object of the data set is attributed to a cluster. By fuzzy clustering, in turn, the grades of membership of an object to all the clusters is extented from the crisp case $\{0,1\}$ to the intervall $[0,1]$. The grade of membership equals to 1 indicates the fact that the object considered belongs to this cluster and 0 excludes its membership.

The set of objects $P = \bigcup_{i=1}^{n} \{P_i\}$ and the set of clusters $C = \bigcup_{j=1}^{n} \{C_j\}$ form a relation, that is a fuzzy one in case of fuzzy clustering. Since the

[1]) Akademie der Wissenschaften der DDR,
Zentralinstitut für Kybernetik und Informationsprozesse, Berlin

universes of discourse are finite the object-cluster-relation on $P \times C$ can be represented by a matrix (cf. Figure 1).

		set of clusters $C = \bigcup_{j=1}^{m} \{C_j\}$					
		C_1	C_2	...	C_j	...	C_m
set of objects $P = \bigcup_{i=1}^{n} \{P_i\}$	P_1	u_{11}	u_{12}	...	u_{1j}	...	u_{1m}
	P_2	u_{21}	u_{22}	...	u_{2j}	...	u_{2m}
	⋮	⋮	⋮		⋮		⋮
	P_i	u_{i1}	u_{i2}	...	u_{ij}	...	u_{im}
	⋮	⋮	⋮		⋮		⋮
	P_n	u_{n1}	u_{n2}	...	u_{nj}	...	u_{nm}

Figure 1 : Object-cluster-relation on $P \times C$
crisp clustering : $u_{ij} \in \{0,1\}$
fuzzy clustering : $u_{ij} \in [0,1]$

There arise two interpretations of this relation. On the one hand a cluster C_j is regarded as fuzzy set on the set of the objects. This corresponds with the column j of the relational matrix.
On the other hand an object P_i can be regarded as a fuzzy set on the set of clusters. That corresponds with the row i of the matrix.
A schematical representation is depicted in Figure 2. Concering this example it has to be taken into account that there is no ordering on the set of clusters.

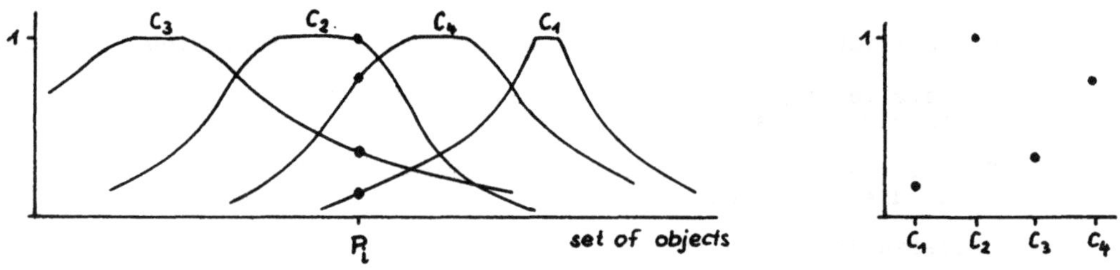

Figure 2 : a) Fuzzy clusters C_1, C_2, C_3, and C_4 on the set of objects P

b) Fuzzy object P_i on the set of clusters C

Because of the fuzzy description of the objects and clusters additional information is delivered as opposed to crisp clustering. Bearing in mind that clustering is mostly a preprocessing tool on the set of objects this information may be used for

- comparing different clustering of the object set achieved by the same cluster technique or by different ones
- composing of clusters
- removing of objects which do not fit well into any cluster

- transferring an object into another cluster.

During a model building process with regard to cluster analysis (cf. also next section) pieces of knowledge of the problem under investigation could not be taken into consideration. Hence, cluster analysis should be applied very carefully.
After having executed the cluster analysis the results should be embedded into the original problem again.
The fuzzy representation of the clusters gives hints how to handle the results of clustering. In this context the 'forgotten' knowledge has to be recalled.

Furthermore, a classificator can be built up by sticking semantics to the cluster which become classes.

Not only disjoint crisp clusters but also non-disjoint ones can be derived from fuzzy clusters. In Figure 3 a schematic example is depicted,

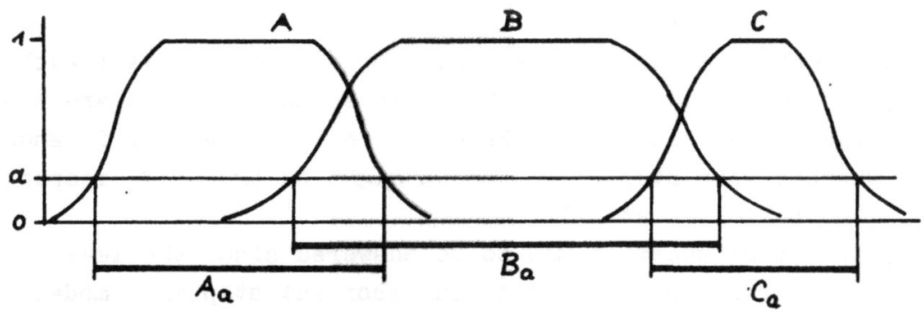

Figure 3: Fuzzy and non-disjoint clusters

where three clusters A, B, and C are described by their membership functions. Let a be a threshold value the depicted non-disjoint clusters (or a-level sets) A_a, B_a, and C_a are assigned.

Another aim of cluster analysis is a compression of data. This can be achieved by not taking a cluster but

- a single representativ
- a few characteristical objects (e.g., lying on the margin of the cluster)
- a parametric representation.

From the methodological point of view, this data compression does not belong to cluster analysis.

2. Model building with regard to cluster analysis

By means of cluster analysis structures are to be found or constructed in a set of objects given.
In doing so, there arise the necessity of a particular model building concerning some problems in applying cluster analysis. Some of these problems shall be discussed.

Before doing cluster analysis at all, the following questions have to be answered:

- What are the objects of the problem in question to be grouped ?
- What about the representation of the objects defined before ?
- How to appraise the similarity between two objects with regard to the properties considered ?

These questions are to be discussed in a more detail.

At the first sight it seems to be trivial to define the objects in the sense of cluster analysis.
In many cases measurements are taken from a real world object such as in technology, biology, medicine etc. Then, these measurements are thought of as the objects for clustering. Before doing so, however, it should be proved whether this step of reasoning might be true with respect to the problem to be solved at all.
Therefore, the first question has to be answered since the level of description is settled here. This is an important step in a model building process.

Very often, objects are represented by features. These features have to carry specific information about the problem in question. In cluster analysis there is always a compensation between the features. Therefore, this property must also be taken into account (cf. discussion of the third question). In general, features are taken from quantities carrying differently (e.g.) physical units. An non-supervised normalization of the features could have catastrophic influences upon the forming of the cluster, as to seen in Figure 4.

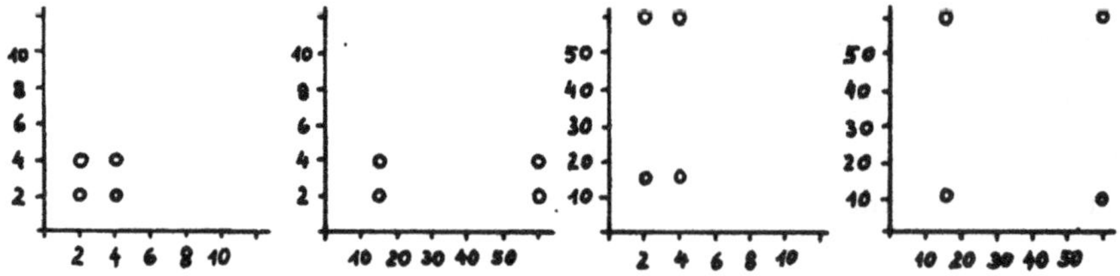

Figure 4: Influence of changing a scale

However, the bottleneck of cluster analysis is to find an appropriate answer of the third question dealing with the assessment of the similarities.

A similarity measure σ has to model the properties of the problem to be structured. According to these properties the objects are to be grouped. Hence, the objects belonging to the same cluster are said to have a common property, that means, they are very similar (see also the first criteria for clustering in section 1).

This must be recalled when a metric is used. In Figure 5 a few unit circles of Minkowski metrics are depicted. The different 'shapes' of

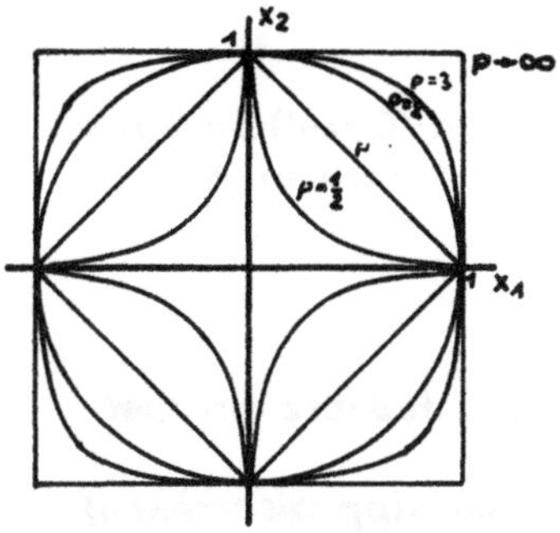

Figure 5: Unit circles $(|x_1|^p+|x_2|^p)^{1/p}=1$ for several $p\in(0,\infty)$

the same numerical value of similarity (distance) are obvious. There arise different compensations between the features for different metrics.

Similarity values $s(P_i,P_k)$ for a pair of objects (P_i,P_k) should be interpretable as follows.

$s(P_i,P_k) = s(P_i,P_l)$	P_i is as similar to P_k as to P_l
$s(P_i,P_k) = s(P_j,P_l)$	P_i is as similar to P_k as P_j to P_l
$s(P_i,P_k) > s(P_i,P_l)$	P_i is more similar to P_k than to P_l
$s(P_i,P_k) > s(P_j,P_l)$	P_i is more similar to P_k than P_j is to P_l

In general, a mapping has to be constructed. It maps a highly dimensional space corresponding to the set of features onto the one-dimensional space of similarity values. Even if the features defined by the findings of the second question involves as much knowledge of the problem as possible such a transformation is always accompanied by loss of information.
As a whole, a proper assessment of similarity values is very crucial.

Some of these difficulties in modelling could be overcome by using k-tupels to characterize similarity. Each entry of the k-tupel can result from

- applying k different similarity measures
- forming k subsets of the features transformed via one or several similarity measures (cf. Figure 6)

However, there arise the problem in processing these k-tupels with a cluster technique that can handle k-tupels of similarity.

Some possibilities for applying common cluster techniques are listed.

- An ordering on the set of these k-tupels can be introduced for a decision making.

- A cluster technique or several different ones are applied k times corresponding to the k entries. After doing so, the k results will commonly be interpreted from different points of view [1].
- The interpretation based on fuzzy description of clusters and entropy of probabilistic sets was introduced in [4].

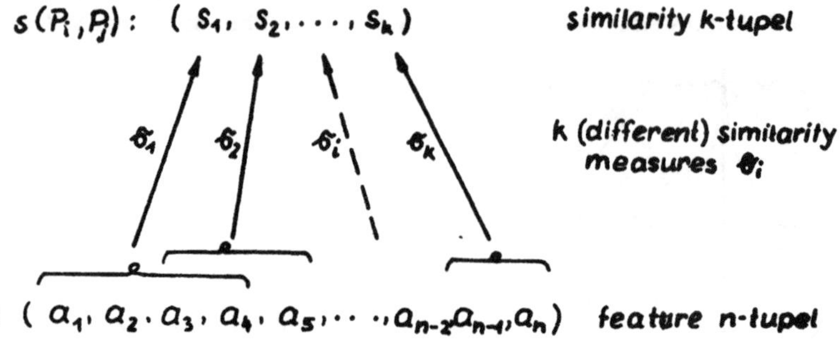

Figure 6: Schematic representation of k-fold appraisement of similarity.

The fact that model building with respect to cluster analyse is a necessary task shall be illustrated by a simple example.

Given the linear differential equation $\ddot{x} + a\dot{x} + bx = 0$ with $x(0)=1$ and $\dot{x}(0)=0$ and a set of pairs of parameters (a,b) (cf. Figure 7). Clustering is to be applied to find out different types of behaviour. Using the euclidian metric the points P_1 is to P_2 more similar (their distance is smaller) than P_1 to P_3 and P_2 to P_4, respectively. Looking at the objects in the parameter space, the points P_1 and P_2 are expected to be grouped into the same cluster (Figure 7).

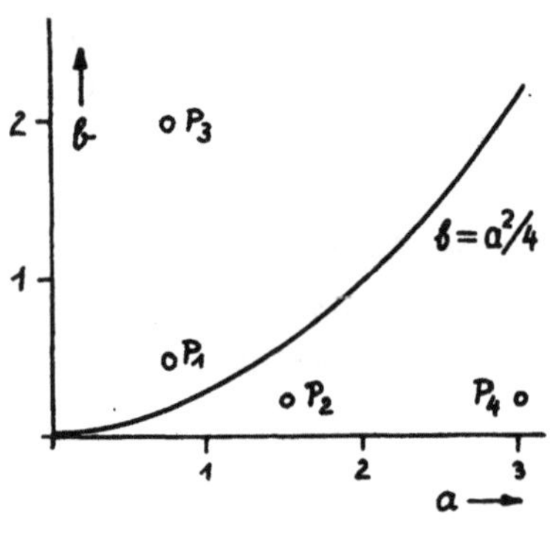

Figure 7: Parameters $P_i=(a,b)_i$ for $\ddot{x} + a\dot{x} + bx = 0$, $x(0)=1$, $\dot{x}(0)=0$

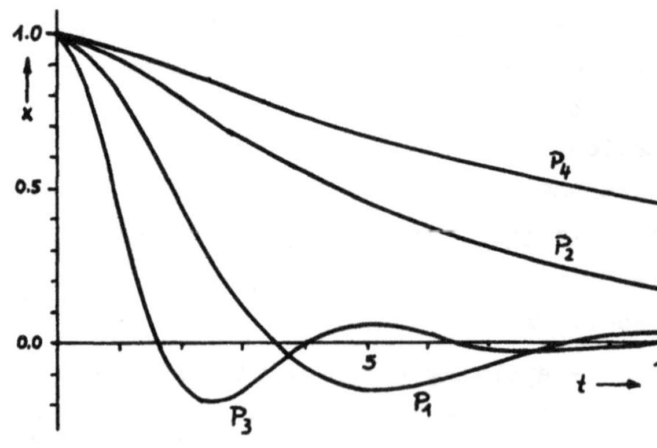

Figure 8: Timefunctions corresponding to the parameters P_i

However, Figure 8 points out that clustering achieves wrong results. Of course, it is well known both the types of behaviour are separated

by the curve $b = a^2/4$.
These differences will appear more clearly with (nonlinear) models having more than two paramaters. Here, the same variation of the values applied to different parameters effects different changes of the solutions of the differential equation.

It has been shown that there need not be a direct line from objects represented by metric features to the application of euclidian metric to calculate the similarity values.
This problem cannot be solved by normalizing the parameters, too.
The question is whether the objects have been defined properly.

As a conclusion, before applying cluster analysis model building according to the three questions has to be done. Furthermore, the cluster algorithm being used or the implementation of such a algorithm have not to restrict the process of modelling.

3. Applications

The author's applications are dealing with structure building or data analysis by means of cluster analysis in the fields of enviromemtal protection, psycho-physiological research, medicine, control.
In this section a few of our applications are outlined.

The decomposition of large-scale systems described by a system of equations into subsystems is discussed in [2]. A 'good' decomposition will have been achieved if both subsystems have as much as possible internal variables and only a few connexions exists between the subsystems.
The objects for clustering are the rows of the structural matrix and the TANIMOTO-metric is used for calculating the similarity values.

This approach has been reported in [3] for optimization of the area of programable logical arrays (PLA). The optimization was based on the decomposition achieved by clustering.

In an ecological system there exists a certain predator-prey-behaviour among the species living there. The structural matrix of this behaviour is depicted in Figure 9.
Applying cluster analysis two large groups are found. It is interesting to note that from the structural point of view the two groups of species are no longer linked if the species No. 4 would die out (cf. Figure 10).
For further investigations the groups were decomposed, in turn.

A similar approach was applied to decompose graph structures (cf. Figure 11). The nodes can be regarded as electrical devices, circuit elements, machines etc. The edges are connexions carrying flows of energy, information, materials etc.
The decomposition was a basis for placement.

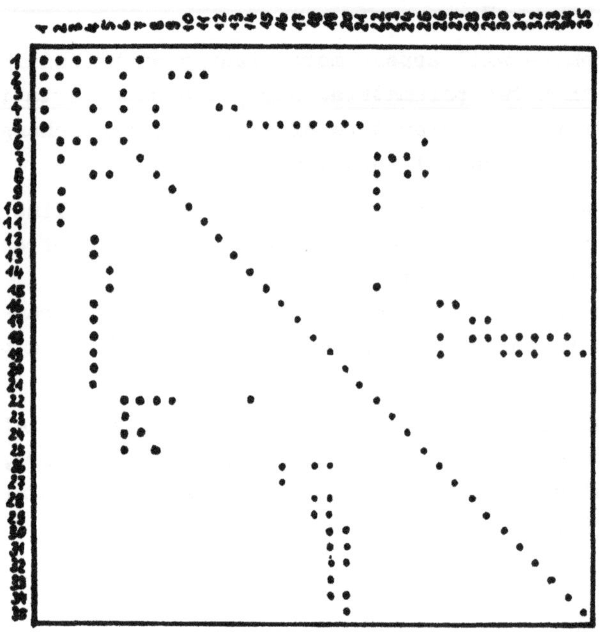

Figure 9: Structural matrix of a predator-prey behaviour in an ecological system

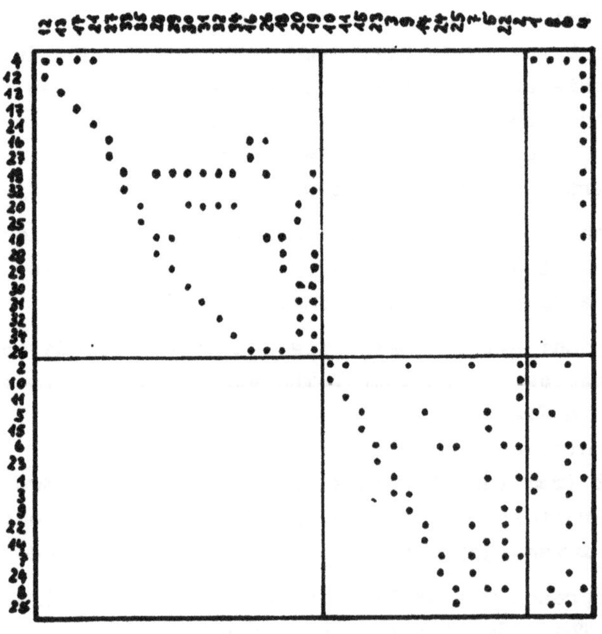

Figure 10: Two substructures within the ecological system of Figure 9

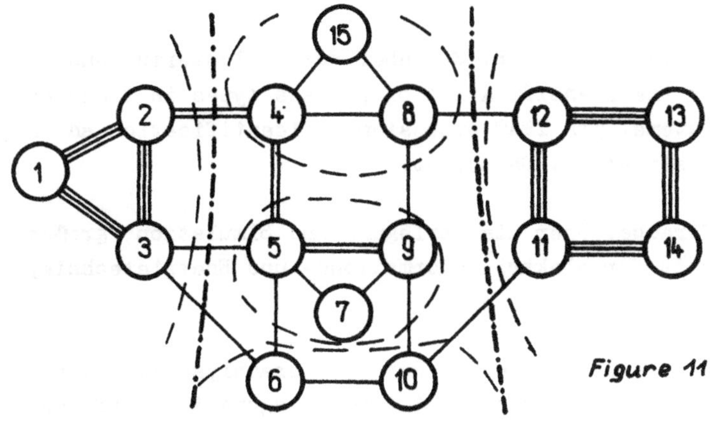

Figure 11: Decomposition of graphs

Finally, an application of cluster analysis in physics will be presented in [6].

A physical model for description of topologically disordered structures is used which is characterized by structural elements corresponding to symmetric cells and their defects. The variation of the density of the different structural elements and their arrangements yield various glassy structures as well as partial and total crystallization within them. Correlations between structural elements result in substructures that are identified by cluster analysis on the basis of geometrical features and internal stress distributions (cf. Figure 12). Estimations for pattern size and distribution of elastic strain energy and other physical parameters are obtained. Cluster regions with high internal stress values can be predominantly transformed into microcells.

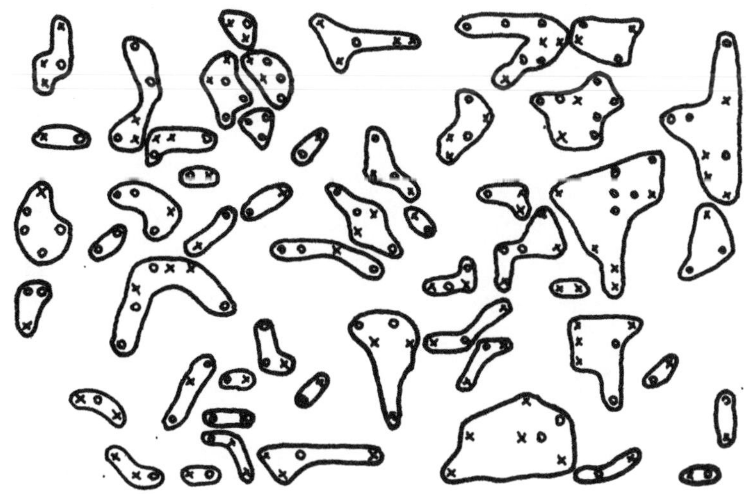

Figure 12: Pattern in a two-dimensional disclination network

4. References

[1] Bocklisch, S.F., Meyer, W. and B. Straube: Fuzzy clustering and classification for the identification of process types in technical and biomedical systems. 5th IFAC symposium on identification and s system parameter estimation. Darmstadt, 1979.

[2] Elst, G. and B. Straube: Über die Zerlegung und Simulation großer Systeme. Zeitschrift für elektr. Informations- und Energietechnik, 11 (1981) pp.129-140.

[3] Franke, G. and B. Straube: Anwendung der Clusteranalyse zur Flächenoptimierung bei PLA-Realisierungen. 11.Arbeitstagung 'Entwurf von Schaltsystemen', Berlin 1982.

[4] Hirota, K. and W. Pedrycz: Characterization of fuzzy clustering algorithms in terms of entropy of probabilistic sets. Pattern Recognition Letters 2 (1984) 213-216.

[5] Richter, A., Straube, B. and W. Pompe: Structural changes in glasses in the context of cluster analysis. Journal of Non-Crystalline Solids (to appear 1985).

[6] Straube, B.: Fuzzy cluster analysis by means of FDA. (in this volume)

FUZZY CLUSTER ANALYSIS BY MEANS OF **F D A**

Bernd Straube [1])

Abstract. The programs of the FORTRAN program package FDA are outlined.

The programs of FDA [1] can be used for finding or building structures by means of cluster analysis.

They are written in FORTRAN IV and run on SM4 and PDP11 minicomputers. The FDA programs are capable of handling up to 1000 objects and 500 features.

Brief description of the programs

A representation of the interrelationship between the programs is depicted in Figure 1.
The user chooses a name <PN> for his problem concerning structure building. (Non-terminals are indicated by angle brackets '<...>'.) After calling any program of FDA this problem name has to read in via terminal. Relating to the problem under discussion the problem name <PN> is also the file name of all the files created by the FDA programs.

Program CHEAD
reads in the data to be clustered. The set of data has to be written according to the syntax diagram <DATA-FILE> :

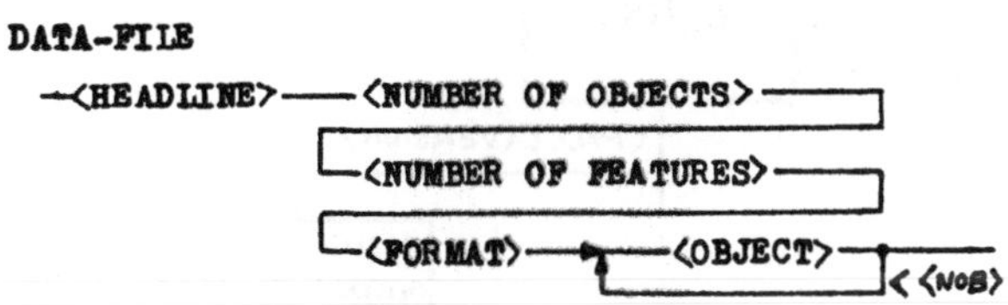

where <OBJECT> is defined as

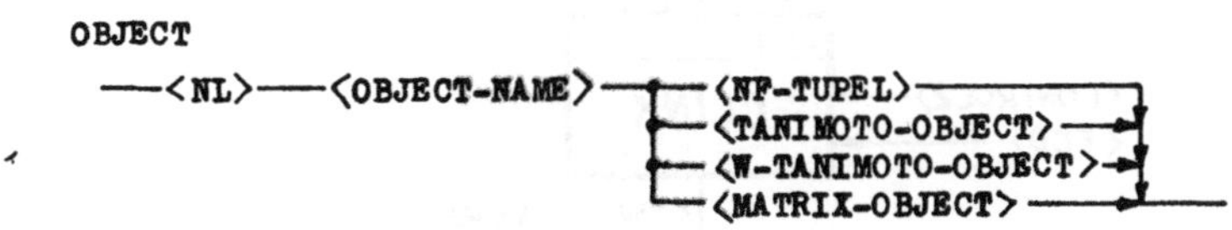

Four examples for <DATA-FILE> are given in Figure 2 .

---

[1]) Akademie der Wissenschaften der DDR
Zentralinstitut für Kybernetik und Informationsprozesse, Berlin

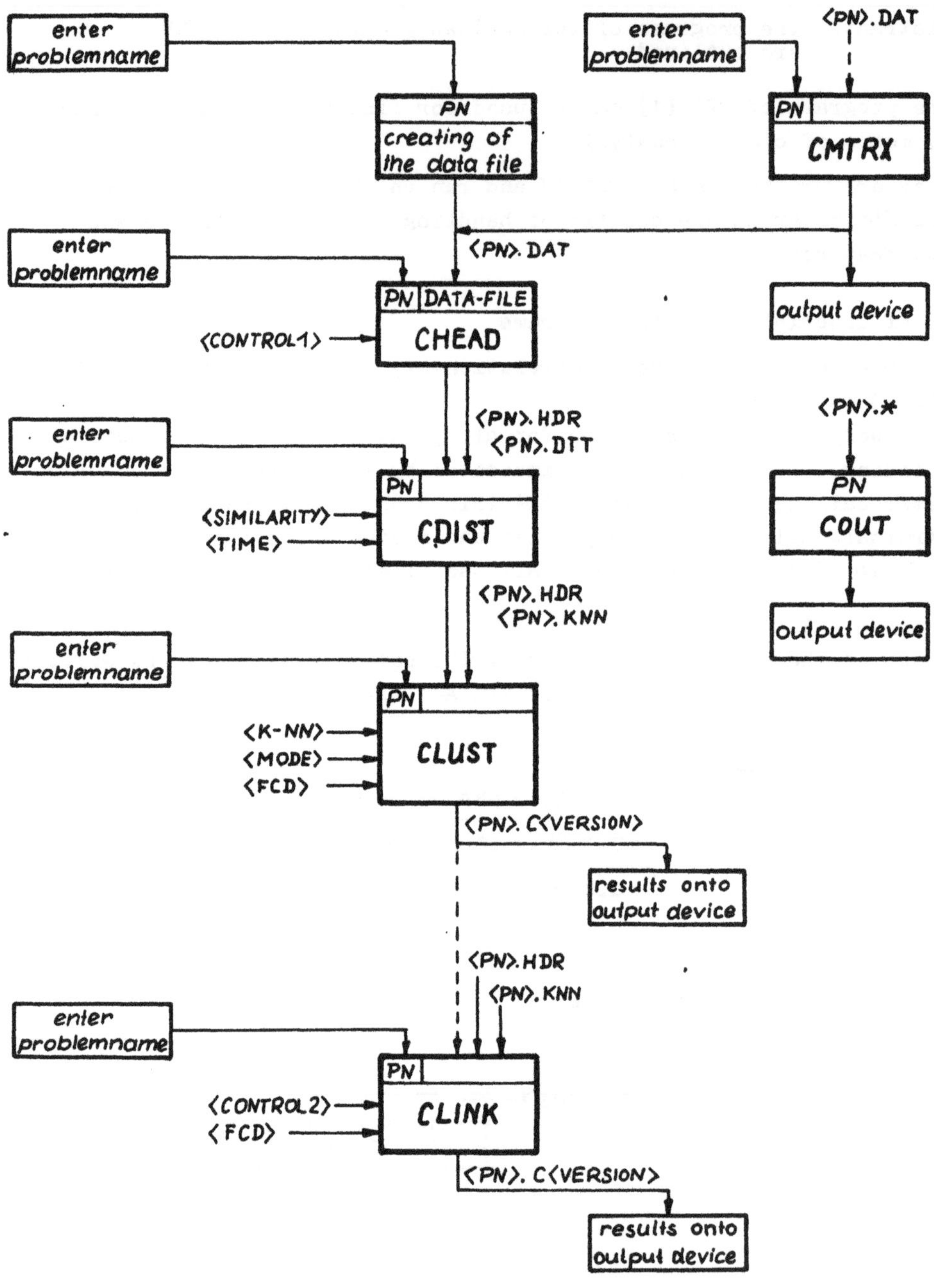

Figure 1 : Schematic representation of FDA

```
/     EXAMPLE 1
/     10 OBJECTS,  2 FEATURES
/     SIMILARITIES VIA EUCLIDIAN METRIC
/     CREATED  JUNE 16, 1982
    NOB=10
    NF=2
(X,A4,X,2(X,F4.1))
OB 1   2.0  3.0
OB 2   3.0  3.0
OB 3   2.0  2.0
OB 4   3.0  2.0
OB 5   8.0 10.0
OB 6   6.0  8.0
OB 7   8.0  8.0
OB 8  10.0  8.0
OB 9   8.0  6.0
OB10  10.0  6.0
```

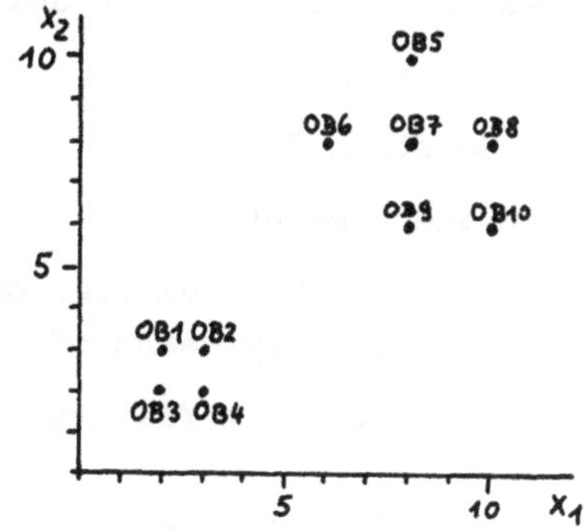


```
/     EXAMPLE 2
/     4 OBJECTS,  6 FEATURES
/     SIMILARITIES VIA TANIMOTO-METRIC
/     CREATED  JUNE 16, 1982
   NOB=4
   NF=6
(X,A4,X,F4.1,X,5(X,F4.1))
R1     3.0   2.0  3.0 10.0
R2     5.0   1.0  2.0  9.0 10.0 11.0
R3     4.0   3.0  4.0  5.0  6.0
R4     4.0   4.0  5.0  6.0  9.0
```

	1	2	3	4	5	6	7	8	9	10	11
R1	·	1	1	·	·	·	·	·	·	1	·
R2	1	1	·	·	·	·	·	·	1	1	1
R3	·	·	1	1	1	1	·	·	·	·	·
R4	·	·	·	1	1	1	·	·	1	·	·

```
/     EXAMPLE 3
/     4 OBJECTS,  12 FEATURES
/     SIMILARITIES VIA WEIGHTED TANIMOTO-METRIC
/     CREATED  JUNE 16, 1982
   NOB=4
   NF=12
(X,A4,2F5.1,X,10(X,2F4.1))
WR01  3.0  3.9   2.0 0.5  3.0 3.1  10.0 0.3
WR02  5.0  7.6   1.0 2.0  2.0 1.5   9.0 0.9  10.0 1.4  11.0 1.8
WR03  4.0 13.3   3.0 2.6  4.0 2.1   5.0 3.0   6.0 5.6
WR04  4.0 13.9   4.0 1.8  5.0 4.1   6.0 6.2   9.0 1.8
```

	1	2	3	4	5	6	7	8	9	10	11
WR01	·	0.5	3.1	·	·	·	·	·	·	0.3	·
WR02	2.0	1.5	·	·	·	·	·	·	0.9	1.4	1.8
WR03	·	·	2.6	2.1	3.0	5.6	·	·	·	·	·
WR04	·	·	·	1.8	4.1	6.2	·	·	1.8	·	·

```
/     EXAMPLE 4
/   5

/     EXAMPLE 4
/     5 OBJECTS,  10 FEATURES
/     SIMILARITIES VIA DISTANCE-MATRIX
/     CREATED  JUNE 16, 1982
   NOB=5
   NF=10
(X,A4,2X,5(2X,A4,F4.1))
OBJA    OBJA 0.0  OBJB 1.1  OBJC 2.1  OBJD 6.3  OBJE 8.0
OBJB    OBJA 1.1  OBJB 0.0  OBJC 3.1  OBJD 7.6  OBJE 6.1
OBJC    OBJA 2.1  OBJB 3.2  OBJC 0.0  OBJD 5.4  OBJE 7.3
OBJD    OBJA 6.3  OBJB 7.6  OBJC 5.4  OBJD 0.0  OBJE 2.0
OBJE    OBJA 8.0  OBJB 6.1  OBJC 7.3  OBJD 2.0  OBJE 0.0
```

	OBJA	OBJB	OBJC	OBJD	OBJE
OBJA	0.0	1.1	2.1	6.3	8.0
OBJB	1.1	0.0	3.2	7.6	6.1
OBJC	2.1	3.2	0.0	5.4	7.3
OBJD	6.3	7.6	5.4	0.0	2.0
OBJE	8.0	6.1	7.3	2.0	0.0

Figure 2: Examples for <DATA-FILE>

Because of the experimental character of cluster analyse CHEAD involves the possibility of changing the data set as follows

- omitting objects
- deleting features

and in the case of metric features

- creating new features or changing the values of a feature by means of
  = scalar o feature --> feature

  or

  = feature o feature --> feature

  where o stands for the usually arithmetic operations +,-,*,/ .

These operations are carried out for all objects via terminal.

Program CMTRX

If data sets are based on

- structural matrices (cf. <TANIMOTO-OBJECT>)

or

- object to object 'distance'-matrices (cf. <MATRIX-OBJECT>)

the program CMTRX can be used for assistant purposes. Via terminal or other input devices the entries of the matrix are read in. The file created by CMTRX fulfills the demands of <DATA-FILE> needed by CHEAD. Optionally, the matrices are printed in their complete form by CMTRX.

Program COUT

Using COUT it is possible to read in all the files created within FDA and to put them onto an output device.
In particular, data files changed by means of the possibilities of CHEAD can put into their topical format in accordance with <DATA-FILE>.
There also exists a two-dimensional plotting of metric features.

Program CDIST

sets up the similarity relation needed for cluster analysis in CLUST.
Following possibilities are available:

- Similarity values are calculated via metrics
  = d_1-metric ($\sum_i |x_i - y_i|$)
  = d_2-metric ($\sum_i |x_i - y_i|^2)^{1/2}$ [euclidian metric]
  = d_∞-metric ($\max_i |x_i - y_i|$)
  = Tanimoto - metric [for structural matrices]
  = weighted Tanimoto - metric [for weighted structural matrices]

- Given explicitely the object to object relationship in the form of a 'distance'-matrix.

- The user's similarity measure coded by himself

By setting an amount of computation time it is possible to calculate the similarity relation step by step. This option is very advantageously in the case of handling large data sets.

Program CLUST

The cluster analysis proper is carried out by means of CLUST. For the problem under discussion, this program can be called again and again without repeating CHEAD and CDIST once more.
Choosing different numbers of nearest neighbours, the controlling parameter <K-NN>, several clusterings are possible to do.

The cluster technique which CLUST is based upon is characterized as follows

- The methode is fuzzy both
  in forming the clusters (by interpreting the underlying heuristics by means of fuzzy sets)
  and
  in the possibility of representing fuzzy clusters
- The technique is non-parametric
- The number of clusters is not predetermined
- The arrangements of the objects in the data file is arbitrary
- Fuzzy k-nearest neighbour rule
- Up to 1000 objects and 500 features can be handled

There is the option of presenting the clusters crisp or fuzzy (cf. <FCD> in Figure 1).

Program CLINK

Bearing in mind that cluster analysis is not only the application of an algorithm but also a proper handling of the data set by the user who brings in his further knowledge of problem to be structured. For this purpose CLINK should be applied.

The results of clustering achieved by CLUST are read into CLINK. Basing on the similarity relation established by CDIST the clusters can be changed. In order to do this the following measures are feasible:

- union of clusters
- taking an object out of a cluster with the aim of
  = putting it into another cluster
  = creating a new cluster. This measure can also be interpreted as deleting this object

The clusters changed in this way are presented in the same way as CLUST does (fuzzy or crisp).

The necessary information for these measures may also be drawn out of the fuzzy description of the clusters and objects.
The membership values will give hints which of the clusters could be joint (for instance, a certain number of clusters is wanted) and/or which of the objects does not fit well.

References

[1] Straube, B.: Anwendungen der Theorie unscharfer Mengen bei der Modellierung realer Systeme, Anhang C. (Applications of fuzzy set theory in modelling real world systems, Appendix C), Dissertation B, Berlin 1984.

FUZZY STRUCTURE FORMATION BASED ON NATURAL SELECTION

Hans-Michael Voigt[1)]

Introduction

One of the most impressive principles of selforganization of complex systems is given by the principle of natural selction as it was presented by Charles Darwin by his fundamental work "on the origin of species by means of natural selection" and its extensions.

Sim plified, this principle explains the evolution of a system to higher stages with respect to its function and structure by mutation and selection.

The question arises how to use this principle for the structural design of complex technical systems efficiently.

For the layout design of printed circuit boards and integrated circuits we want to try to answer this question.

Replication

To speak of mutation and selction in a general sense we use the notion of "replicators" as it was introduced by Dawkins.

By "replicators" we mean any units which may be copied and

(i) which imply by their nature a certain influence on the probability or possibility to be copied

(ii) which may be, at least principally, ancestors of an indefinitely long series of copies.

Replicators in this sense may be alleles in a gene pool, relative numbers within animal populations, selfreproducing macromolecules, strategies of individuals etc.

The replication of these units will be modeled by the following general replicator equation

$$\frac{\dot{x}_i}{x_i} = f_i(x_1,\dots,x_n) - \phi \qquad (*)$$

or

$$\frac{x_i'}{x_i} = \frac{f_i(x_1,\dots,x_n)}{\phi} \qquad (**)$$

with $\sum x_i = 1.$

1) Academy of Sciences of the G.D.R., Mathematics and Informatics Division, DDR-1199 Berlin-Adlershof, Rudower Chaussee 5

In terms of genetics x_i is the proportion of allel i within the gene pool, f_i is the mean fitness of allel i within the gene pool, and $\emptyset$ is the mean fitness of the total gene pool. If we include the possibility of mutations by determined mutation rates this yields a model for natural selction processes.

Obviously stable structures of the evolving system are given by stable attractors - in the most simple case by stable fixed points $\dot{x}_i = 0$ and $x_i' = x_i$, resp. - which represent a stationary probability or possibility distribution x_i of allel i within the gene pool. This dynamic evolution of stationary structures is driven by competition and cooperation between the alleles.

Linear fitness functions

Of special interest are linear homogenous fitness functions f_i.
In this case we have

$$f_i(x_1,\ldots,x_n) = \sum_j G_{ij} x_j = (Gx)_i$$

and

$$\emptyset = \sum_i x_i \sum_j G_{ij} x_j = \sum_i x_i (Gx)_i = (x,Gx).$$

The G_{ij} are interaction coefficients between alleles i and alleles j. The replicator equation with linear fitness functions has the following properties, which are important for the further analysis:

(i) Eq. (*) is invariant under the transformation $G_{ij} := G_{ij} + a_j$

(ii) Eq. (**) is invariant with respect to the fixed points under the transformation $G_{ij} := G_{ij} + a_j$

(iii) For G real, symmetric and $G_{ij} \geqslant 0$ the following relations for the mean fitness of the total gene pool hold $\dot{\emptyset} \geqslant 0$ (*) and $\emptyset' \geqslant \emptyset$ (**).

(iv) For G real, symmetric and $G_{ij} \geqslant 0$ the eigenvalues of eq. (*) are real, i.e. no limit cycles or other dynamic attractors occur.

Replication rate control

For the control of the structure formation process we introduce variable selfreplication rates $G_{ii} = u_i$.
By means of these replication rates a scalarization of the global state of structuring is possible. $u_i \to -\infty$ yields a uniform cooperation between the replicators and $u_i \to \infty$ a uniform competition. Between these two extremal cases further states of structuring occurs. Especially interesting are homogenous selfreplication rates $u_i = u$.

Partitioning of graphes

The partitioning of graphs and hypergraphs, resp., is one of the problems which has to be solved in the design of printed circuit boards and integrated circuits. This problem gets more and more difficult to solve because of the increasing complexity of the integrated circuits to be designed. Therefore combinatorial methods are prohibitive. As a suitable method to overcome these difficulties the use of the principle of natural selection seems to be an effective heuristics.

The problem of graph partitioning may be stated as follows: Given is an undirected graph $G = (V,E)$ with the edge set E and the vertex set V. The edges have the weights $G_{ij} = G_{ji}$, $G_{ij} \geqslant 0$. The vertices should be assigned to partitions of the partition set P. With x_{ik} the membership of vertex V_i to the partition P_k is denoted. Then the partitioning matrix C is given by

$$C = (X, GX)$$

with $X = [x_{ik}]$ the matrix of membership values. Here C_{kk} is the fuzzy number of weighted edges within the partition P_k and C_{kl}, $k \neq l$, the fuzzy number of weighted edges between the partitions P_k and P_l.

The partitioning problem of graphs can then be formulated as a vector optimization problem with

$$C_{kl} \rightarrow \max \quad \text{for} \quad k = l$$

or equivalently

$$C_{kl} \rightarrow \min \quad \text{for} \quad k \neq l.$$

From this a number of special cases based on the used utility function may be derived where the following are especially interesting for the above given partitioning problem

$$\sum_k \; l_k \, C_{kk} \rightarrow \max \qquad (+)$$

$$\prod_k \; C_{kk}^{\,l_k} \rightarrow \max \qquad (++).$$

Eq. (++) means for $l_k = 1$, that the C_{kk} should be maximized and should have at the same time approximately equal size. This case occurs in the design of integrated circuits based on standard cells and gate arrays.

Diversified replication

Related to the previous sections we have for every partition P_k a replicator equation of the form

$$\frac{\dot{x}_{ik}}{x_{ik}} = \frac{(Gx_{.k})_i}{C_{kk}} - e_i \qquad (!)$$

or

$$\frac{x'_{ik}}{x_{ik}} = \frac{(Gx_{.k})_i}{C_{kk}\, e_i} \qquad (!!)$$

with $\sum_k \; x_{ik} = 1.$

Corresponding to the properties of the replicator equation (*) and (**) eqs. (!) and (!!) have the following properties:

(i) Eqs. (!) and (!!) are invariant with respect to the fixed points under the transformation $G_{ij}:=G_{ij}+a_j$
This property will be used for negative replication rate control $u < 0$ such that $G_{ij} \geqslant 0$.

(ii) In analogy to property (iv) of the replicator equation (*) for eq. (!) no limit cycles or other dynamic attractors occur. This property ensures a convergence of the diversified replicator equation.

(iii) For G real, symmetric and $G_{ij} \geqslant 0$ corresponding to property (iii) of the replicator equation (*) for the diversified replicator equation (!) the following relation holds

$$(\prod_k \; c_{kk})^{\cdot} \geqslant 0 \; .$$

The same holds for the eq. (!!) with

$$\prod_k \; c'_{kk} \geqslant \prod_k \; c_{kk} \; .$$

The equality is valid for the fixed points of eqs. (!),(!!).

Eq. (!!) extented by a mutation rate control m

$$x'_{ik} = \left(x_{ik} \; \frac{(Gx_{.k})_i}{C_{kk}} + m \right) / \, e_i \qquad (!!!)$$

was implemented on a home computer robotron Z 9001 as a method for graph partitioning.

Some results for an industrial example are shown in figs. 1,2,3.

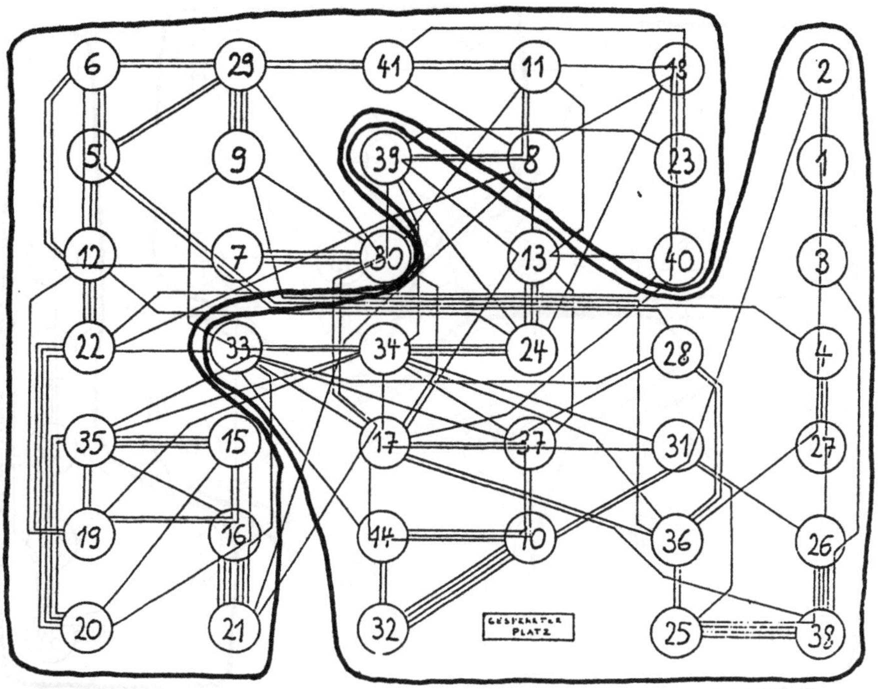

Fig. 1 2 partitions, control sequence u = -1.0, -0.9, -0.8, -0.5, -0.2, partition size 20/21, 20 cuts, 172 iterations

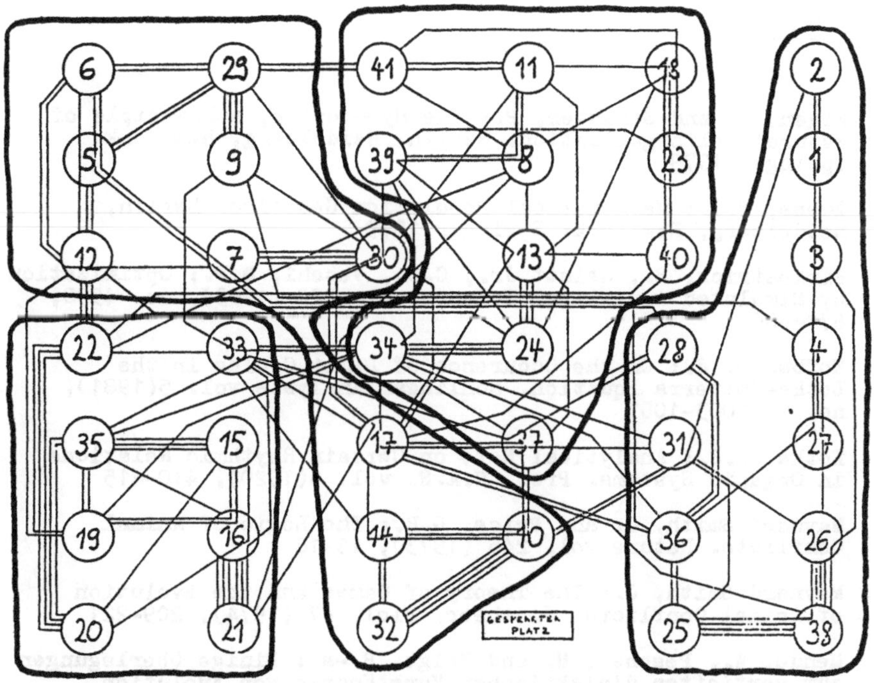

Fig. 2 4 partitions, control sequence u = -1.0, -0.5, -0.3, -0.2, -0.1, partition size 11/11/11/8, 38 cuts, 229 iterations

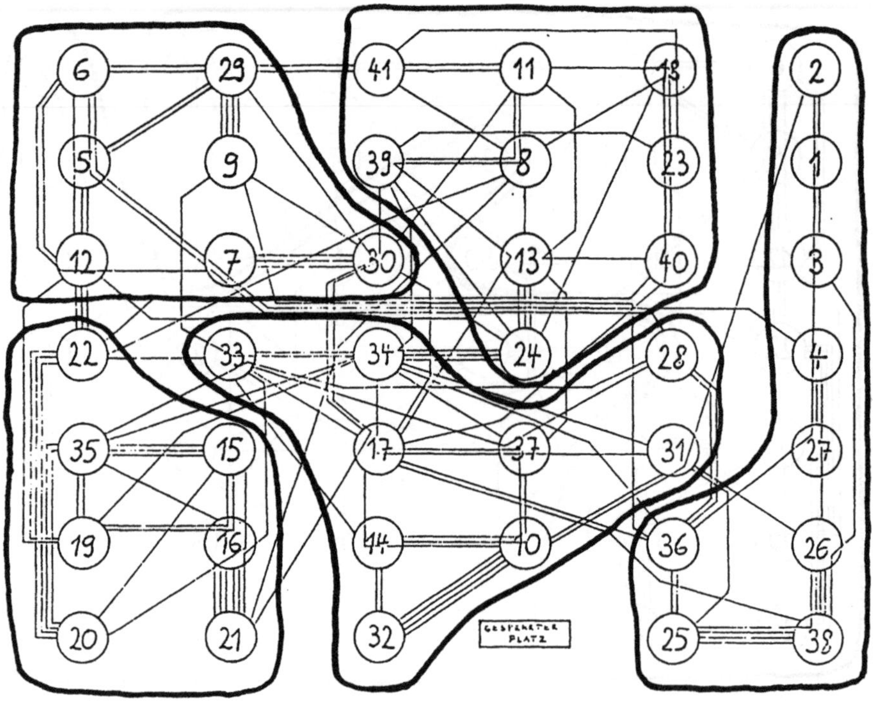

Fig. 3 5 partions, control sequence u = -4.0, -3.0, -2.0, -1.0,-0.5, partition size 9/9/9/7/7, 39 cuts, 165 iterations

References

[1] Eigen, M. and Schuster, P.: The Hypercycle, a Principle of Natural Selforganization. Berlin, Heidelberg, New York: Springer 1979

[2] Ewens, W.J.: Mathematical Population Genetics. Berlin, Heidelberg, New York: Springer 1979

[3] Kirkpatrick, S., Gelatt Jr., C.D., Vecchi, M.P.: Optimization by Simulated Annealing. Science vol. 220 (1983), no. 4598, 671-680

[4] Hofbauer, J.: On the Occurence of Limit Cycles in the Lotka-Volterra Equation. Nonlinear Analysis vol. 5(1981), no. 9, 1003-1007

[5] Lotka, A.J.: Analytical Note on Certain Rhythmic Relations in Organic Systems. Proc. N.A.S. vol. 6(1920), 410-415

[6] Maynard Smith, J. and Price, G.R.: The Logic of Animal Conflicts. Nature vol. 246 (1973), 15-18

[7] Maynard Smith, J.: The Theory of Games and the Evolution of Animal Conflicts. J. theor. Biol. 47 (1974), 209-221

[8] Mende, W., Peschel, M. and Voigt, H.-M.: Einige Überlegungen zum vertieften dialektischen Verständnis von Evolution. Sitzungsberichte der Akademie der Wissenschaften der DDR no 10G, (1983), 65-74

[9] Peschel, M. and Mende, W.: Leben wir in einer Volterra-Welt? Ein ökologischer Zugang zur angewandten Systemanalyse. Berlin: Akademie-Verlag 1983

[10] Peschel, M., Mende, W., Richardt, J. and Voigt, H.-M.: Solution of Nonlinear Assignment Problems and Structure-Building Procedures on the Base of the Fuzzy Approach. Sitzungsberichte der Akademie der Wissenschaften der DDR no. 10N (1983), 10-21

[11] Schuster, P., Sigmund, K., Hofbauer, J., Wolff, R., Gottlieb, R. and Merz, P.: Selfregulation of Behaviour in Animal Societies. Biol. Cybernetics 40(1981), 1-25

[12] Taylor, P. and Jonker, L.: Evolutionary Stable Strategies and Game Dynamics. Math. Biosci. 40(1978), 145-156

[13] Voigt, H.-M.: Evolution Processes and Structuring. In: Report "Tagung Numerische Lösung von Differentialgleichungen 1982". Institut für Mathematik der AdW der DDR, R-MATH-01/83, 242-249

[14] Voigt, H.-M.: Structuring and Evolution Processes. In: Proceedings of the 5th Meeting of the UNESCO Working Group on Systems Theory 1984. Berlin: Akademie-Verlag (to appear)

[15] Volterra, V.: Lecons sur la theorie mathematique de la lutte pour la vie. Paris: Gauthiers-Villars 1931

[16] Zeeman, E.D.: Population Dynamics from Game Theory. In: Global Theory of Dynamical Systems. Lecture Notes vol. 819. Berlin, Heidelberg, New York: Springer 1980

[17] Hofbauer, J. and Sigmund, K.: Evolutionstheorie und dynamische Systeme. Berlin, Hamburg: Paul Parey 1984

[18] Voigt, H.-M.: Structure formation processes, optimization and evolution: An ecological approach. In: Wissenschaftliche Zeitschrift der TH Ilmenau (to appear)

Acknowledgement

The author would like to thank Professor M. Peschel, head of the Mathematics and Informatics Division of the Academy of Sciences of the G.D.R., for his steady stimulation during the development of this work. Thanks are due to Dr. W. Jansen, Central Institute of Cybernetics and Information Processes of the Academy of Sciences of the G.D.R., for the cooperation in proving some properties of the diversified replicator equation.

AN APPLICATION OF A FUZZY PETRI NET IN COMPLEX INDUSTRIAL SYSTEMS

H.-P. Lipp[1)]/R. Guenther[2)]

Introduction

Complex industrial processes, such as chemical reactors, pulp mills, cement kilns or glassmacking cannot be satisfactorily controlled, using traditional methods of the control theory. These processes are characterized by

- many sub-systems, which are connected by flows of workpieces, materials, energy or information
- temporal varibility of structure and functions
- "freedom of will" of the people, working in it and
- the conflicting goals of the different sub-systems.

At the operative control in this system the regime of technology is coordinated in the sub-systems by the human operator, to reach the production goal as efficiently as possible. In this multi-decision making process the operator applies an incomplete plant model, in which disturbances and the effects of control actions are recognized quickly. The incomplete plant model is supported by fuzzy process knowledges, resulting from the operator's subjective experiences. The fuzziness results in this systems from

- the unknown plant model
- the simplification of complex models to such models, which enable a effective selection of control actions
- the subjective valuation of complex process variables, such as the drop shap of molten glass in glassmacking
- the unknown process goal, when it results from the compromise of many sub-goals.

In order to include the operator's process experiences in an automation schema, a fuzzy Petri net is introduced. In this paper we briefly present at first modelling of complex processes by a classical Petri net, and then we describe the main ideas of the fuzzy Petri net, pointing out some specific features. This fuzzy Petri net is supported by the operator's subjective experiences and is applied as a formal model to complex processes. At last we apply it to a control system in a pulp mill.

---

1) Technical University Karl-Marx-Stadt

2) Pulp and paper mill Rosenthal

Modelling of complex processes by classical Petri net

There are many ways to represent the behaviour of a complex system by a formal model. In particular such formal descriptions of dynamic systems are necessary, if the system contains interacting components that are working in parallel. Then by the unknown relative speed of these concurrent actions unexpected effects may appear. By the complexity of systems such undesired effects will increase. The Petri net model solves these problems by the following important characteristics:

- Independent actions are represented independently in the model by such components, on which they depend and where change will occur.
- Only the flow of control is modelled.
- All important properties of the model are given by a graph.

These properties are more orless common to all kinds of models. The classical Petri net consists of places and transitions, are connected by directed edges.

A Petri net is defined by

$$NP = [P, T, F, C, V, K, M, m_0] \tag{1}$$

$P = \{p\}$	is a finite set of places
$T = \{t\}$	is a finite set of transitions
F	is a transition function $F : (P \times T) \cup (T \times P) \longrightarrow \{0, 1\}$
$C = \{c\}$	is a finite set of colours
V	is the multiplicity of tokens defined on an arc $V : (F \times C) \longrightarrow IN$
K	is the token capacity in the places $K : (P \times C) \longrightarrow IN$
M	is the set of allowable markings $M : (P \times C) \longrightarrow IN$
$m_0 \in M$	is the initial marking of the net.

The analogy between the Petri net and the components of a complex system can be exemplified by the following table:

Petri net		plant
transitions	⟶	actions, sub-systems, machines
places	⟶	conditions for ability of work in the separate sub-systems (for instance buffers, tanks, store or tolerances)
token flows	⟶	mass flows, flow of workpieces, energy or materials
arcs	⟶	structures of the plant

The different colours serve for the simplification of the plant model, when in different situations must be switched between several models. A place may have one or many tokens or is may be empty. When there is one token or many tokens in each input place of a transition according to the capacity, then this transition is ready to fire. The act of firing involves removing tokens from each input place and putting in each output place according to the multiplicity of the arcs. The places indicate the holding or not holding of a condition in the process model and transitions that describe the event of some conditions represent the control actions. Usually a Petri net is represented by a bipartibe graph, graphically we use circles for places and bars for transitions.

The disadvantage for this classical Petri net consists in the "sharp"-behaviour of places and transitions. Also coloured Petri nets can switch between idealized model components only. In many real processes a transition is not an action only, but a set of actions, for instance, actions for normal, overload, and underload situations. The conditions for ability of work in a sub-system of a real plant cannot be represented by a precise marking in some places, but by a set of markings in those components. In many real processes the different control goals and the constraints on the choice between different control alternatives are fuzzy. They cannot be represented by the classical Petri net. In order to include this fuzziness of real process in a plant model, a fuzzy Petri net is introduced.

Representation of the fuzzy Petri net

The fuzzy Petri net is derived from the classical Petri net by fuzzification of places and transitions.

A fuzzy Petri net $\widetilde{NP}$ is defined by

$$\widetilde{NP} = (\widetilde{P}, \widetilde{T}, \sigma, \pi, \eta, M, m_0) \qquad (2)$$

$\widetilde{P} = \{\widetilde{p}\}$	is a finite set of fuzzy places $\widetilde{p} = \{(p; \mu_{\widetilde{p}}(p))\}$
$\widetilde{T} = \{\widetilde{t}\}$	is a finite set of fuzzy transitions $\widetilde{t} = \{(t; \mu_{\widetilde{t}}(t))\}$
σ	is a transition function $\sigma: (\widetilde{P} \times \widetilde{T}) \cup (\widetilde{T} \times \widetilde{P}) \longrightarrow [0, 1]$
π	is a start function $\pi: (\widetilde{P} \times \widetilde{T}) \longrightarrow [0, 1]$
η	is a final function $\eta: (\widetilde{T} \times \widetilde{P}) \longrightarrow [0, 1]$
M	is the set of markings in $\widetilde{PN}$
$m_0 \in M$	is the initial marking of $\widetilde{PN}$

This fuzzy Petri net $\widetilde{PN}$ is a concentration of many classical Petri nets, which result from different models of a real process in different situations. The places and transitions are fuzzy sets, characterizing the variation of constraints or control actions in the real process. Their membership functions and the fuzzy valuation of the start, final and transition functions are based on intuitions and experiences of human operators. These membership functions can also be used for modelling the behaviour of the humans in the plant.

A fuzzy place contains many tokens and stands for all conditions of a control action in the different process situations. Their membership function describes the goodness for firing transitions (see Fig. 1). Many or a few tokens in a place mark disturbances, which restrict the safety of the network. The membership function $\mu_{\tilde{p}}$ is a fuzzy measure for the capacity of places, and the number of tokens in a place is an expression for the ability to work in the adjoining transitions.

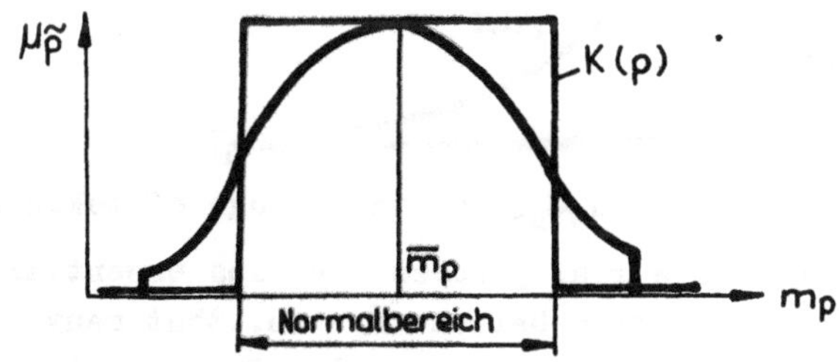

Fig. 1: Membership function $\mu_{\tilde{p}}$ of a fuzzy place

The fuzzy transitions $\tilde{T}$ of the network are a concentration of many "sharp" control actions to fuzzy sets. Their membership functions result from experiences of a human operator and are a measure of the qualification of these "sharp" actions in a special control task. Different actions of a transition $\tilde{t}$ are described by different token flows between the input and output places. The ratio of token flows of a transition $\tilde{t}$ is called the recipe of transition. By the continuous adaption of token flows between the transition the safety in the network is increased for a long time.

A high value for liveliness is described by good membership of applied control actions.

The functions σ, π and η are used for the selection from fuzzy transitions for the improvement of safety and liveliness in the network. The transition function $\sigma_{\tilde{t}}$ describes the dynamics of the transition $\tilde{t} \in \tilde{T}$, which results from the change of its token flows.

The start function $\pi_{\tilde{t}}$ describes the fuzzy ability to fire of the transition $\tilde{t} \in \tilde{T}$ in state place, and the final function represents a fuzzy measure for compromise between the goal of token flows in different transitions and desirable constraints in the places. All these membership functions result from experiences of operators consulting different specialists.

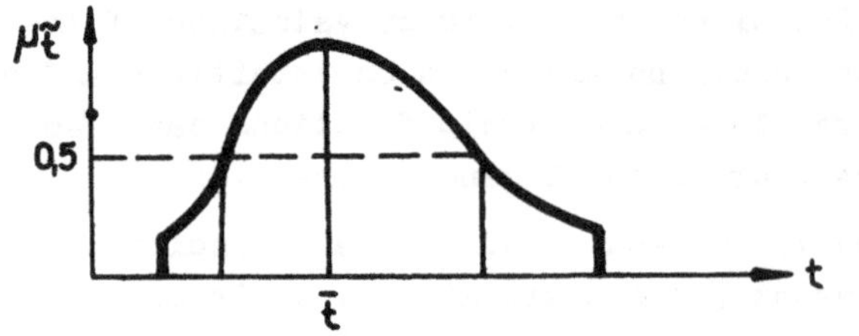

Fig. 2: Membership function of a transition $\tilde{t}$

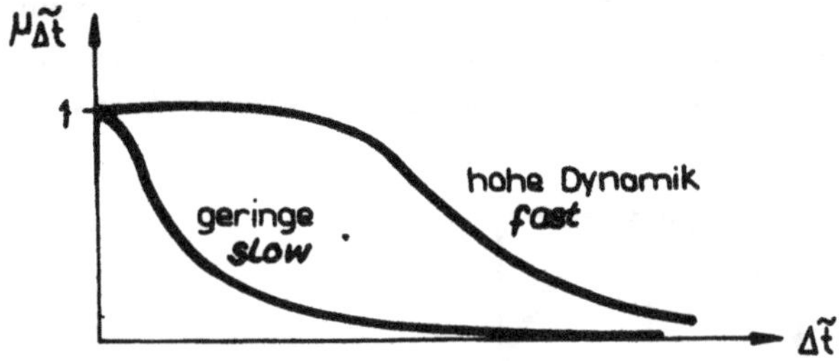

Fig. 3: Membership function $\mu_{\Delta\tilde{t}}$ for the change of token flows

The fuzzy Petri net of a real process contains essentially structur informations in form of many balance spaces, that many different token flow coordination strategies are possible. The changing of coordination conditions enables robust models for difficult process situations.

This fuzzy Petri net can be transformed to a classical Petri net, changing membership functions to sharp memberships. Hence, the modelling of continuous and discontinuous control actions is possible in the same network. The fuzzy Petri net can be deduced from many different classical Petri networks, each for special process situation. As long as the fuzzy sets are finite, the fuzzy model is equivalent to a (much larger) coloured Petri net.

Determination of control policies in the fuzzy Petri net

In this fuzzy Petri net model disturbances are identified by irregular token levels in the places. The elimination of disturbances is possible by coordination of token flows, so that acceptable levels result in the places. The problem of a control system in a fuzzy Petri net consists in selecting the transitions, whose token flows improve the level in the places. A sequence of "sharp" control actions $\underline{t}_0$, $\underline{t}_1$, $\underline{t}_2$, ..., $\underline{t}_m$ is called a control policy, when the initial marking m_0 is transformed to the final marking and the liveliness of the network is guaranteed. At first this policy is determined by selecting a sequence from fuzzy transitions $\underline{\tilde{t}}_0, \underline{\tilde{t}}_1, \underline{\tilde{t}}_2, \ldots, \underline{\tilde{t}}_k$, which then is transformed to a sequence of "sharp" transitions $\underline{t}_0, \underline{t}_1, \underline{t}_2, \ldots, \underline{t}_m$. Consequently the "sharp" transitions $\underline{t}_t$ represent the realizations of the fuzzy transitions $\underline{\tilde{t}}_t$ at the time t. Their determination is performed by the adaption of fuzzy transitions to each other, so that the liveliness and the safety of the network is as much as possible.

When a control strategy must working for a long time, the transitions are selected in a bottom up mode, where at first good token levels are realized in the places of the network and then, when the ability of work in all places are very good, the goal of token flows in the transitions is realized. But when the control strategy is working for a short time, then the fuzzy transitions are selected in a top down mode. In this mode at first the goal of tokenflows is realized and then good token levels in places are aspired. In generally in the determination of control actions in the fuzzy Petri net the top down and the bottom up mode are applied simultaneous.

The changing of token flows in the fuzzy Petri net is determined in the following steps:

- Selection of the fuzzy transitions $\tilde{T}_{MIN}$ according to the most effective place $\tilde{p}_{MIN}$ for the determination of the control strategy. This place can represented by the place with the "badest" membership value

$$\mu_{\tilde{p}_{MIN}} = \underset{\tilde{p} \in \tilde{P}}{MIN} \left(\mu_{\tilde{p}}(p) \right) \tag{3}$$

$$\tilde{T}_{MIN} = \left\{ \tilde{t}_{\tilde{p}_{MIN}} \right\} \text{ are all } \tilde{t} \text{ adjoining to } \tilde{p}_{MIN}$$

- Selection of the fuzzy transitions $\tilde{T}_D \subset \tilde{T}_{MIN}$ according to the most effective control direction for compensating disturbances (see Fig. 4)

$$\mu_{\tilde{T}_D} = \underset{\tilde{t} \in \tilde{T}_{MIN}}{MAX} \left(\mu_{C_{\tilde{t}}}(\bar{t}) \right) \tag{4}$$

- Selection of a "sharp" transition $\tilde{t}_t \in \tilde{T}_D$ according to the best liveliness of the fuzzy transitions

$$\mu_{\tilde{t}_t}(t) = \underset{t \in \tilde{T}_D}{MAX\ MIN} \left[\pi_{\tilde{t}_t};\ \delta_{\tilde{t}_t};\ \eta_{\tilde{t}_t} \right] \tag{5}$$

All realizations t of the fuzzy transitions $\tilde{t}$ are determined in an iterativ process in this way till all places $\tilde{p}$ and fuzzy transitions $\tilde{t}$ can not be further improved.

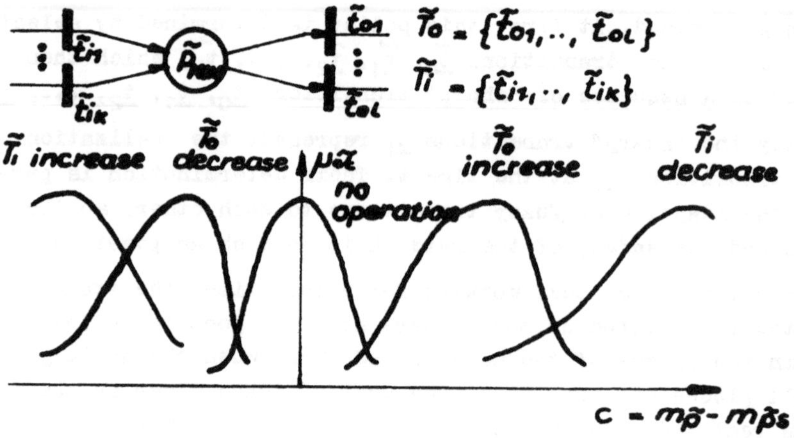

Fig. 4: Membership functions for selecting fuzzy transitions $\tilde{t}$ in place $\tilde{p}$

Application of the fuzzy Petri net for a control system in a pulp mill

The control of a pulp mill is very complicated, because this plant is a very complex system with large time constants. A real-time computer is used for process control and production planning, which will essentially coordinate the different sub-processes in the plant. The technological line in the pulp mill is a very complicated dynamic system, which must be approximated by a simple model. The main elements of the line are tanks and sub-systems, which are connected by mass flows, such as pulp, chemicals, liquor, steam or energy flows. The problem of control is formulated by calculation of production schedules, involving

- mass flows coordination for the elimination of disturbances
- mass flows coordination in repairing situations
- changing in the production rates of the process
- possibility of indirect storage of steam
- acceptable tank levels for a stable production process at the end of the planning period.

The fuzzy Petri net model of the pulp mill consists of 15 transitions and 20 places, which are connected by 46 directed arcs (see Fig. 5). In this model the places stand for tanks or buffer, and the transitions represent sub-processes, such as digester, bleach plant, paper mill and so on. The token flows of the arcs describe the structural dependences between the sub-systems, such as pulp, liquor, chemical and steam flows. The ratio between token flows around a transition is built in an updating process and is dependent on the production rate or the quality of product.

If a temporary limitation of the desired production rate is caused by disturbances, then the level in the tanks represents a measure of process stability. When the limitation is removed the dispatcher system must coordinate the process units, that the system works in the previous optimal steady state for a long time.

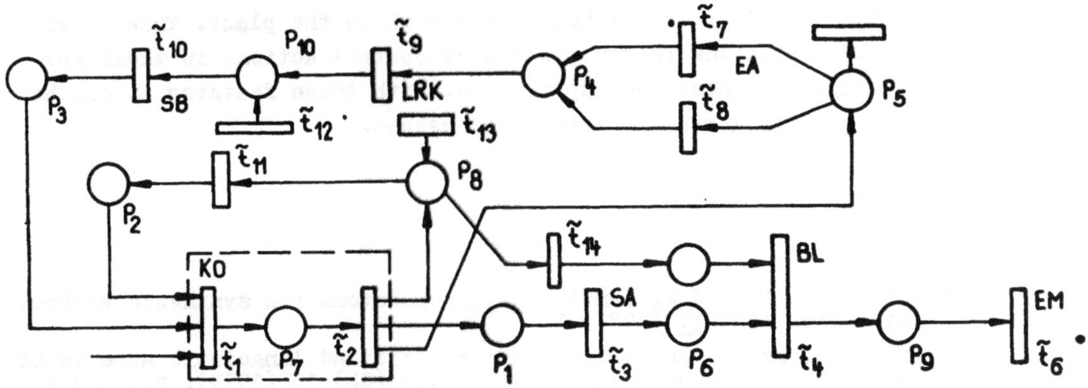

Fig. 5: A fuzzy Petri net model of a pulp plant

In dialouge with the dispatcher the computer calculates production schedules, involving for tank levels and mass flows in time steps of 30 minutes for a process time in 24 hours. This strategy protocol is calculeted in real-time processing by computer in a short cycle. It describes the efficiency of control actions and disturbances and helps the dispatcher in his decision process.

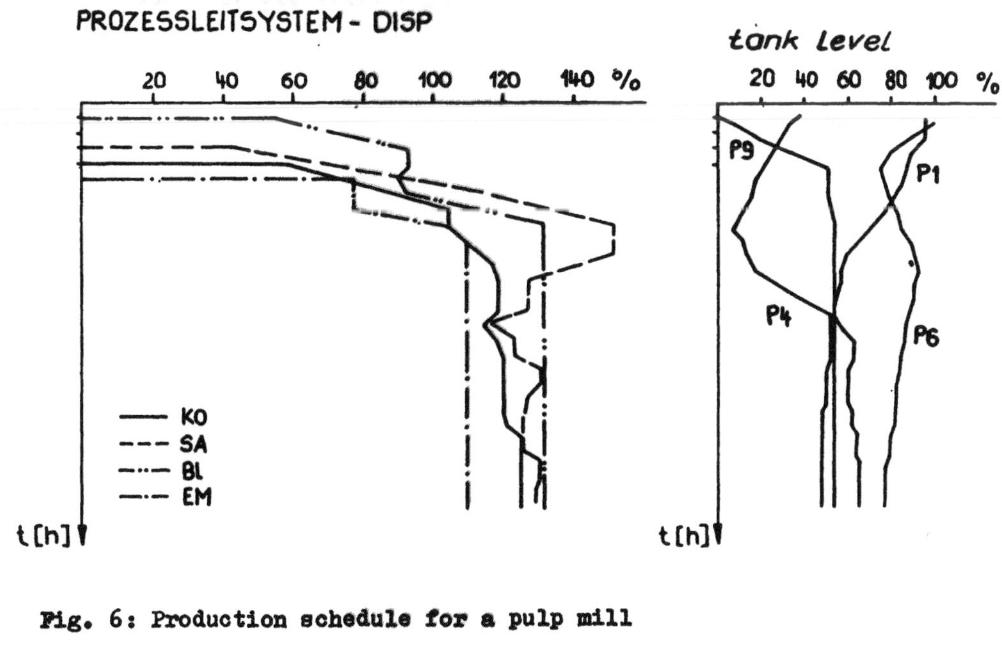

Fig. 6: Production schedule for a pulp mill

Conclusion

In this paper a fuzzy Petri net for modelling a complex process was presented. This fuzzy Petri net was derived from the classical Petri net by fuzzyfication of places and transitions. The fuzzy membership functions result from subjective experiences of the operator and describes the behaviour of the humans who work in the plant. This Petri net model permits the representation of control actions in their normal, overload or underload application. With these features it can be used for modelling real industrial processes.

References

[1] Benlahcen, D. and M. Lamotte: A fuzzy automation synthesis Method. Automatica (GB) 17 (1981) 2.

[2] Braae, M. and D. Rutherford: Theoretical and linguistic aspects of the fuzzy-logic-controller. Automatica (GB) 15 (1979), pp. 523 - 527.

[3] Genrich, H. E. and K. Lautenbach: Net theory and applications. Lectures Notes in Computer Sci. (1979) 84, Springer, Berlin.

[4] Gluss, B.: Fuzzy multi-stage-decision-making, fuzzy state and terminal regulators and their relationsship to non-fuzzy quadratic state and terminal regulators. Int. J. Control 17 (1973) 1, pp. 177 - 192.

[5] Lipp, H.-P.: Application of fuzzy Petri nets for description of control policies in complex production systems. Wiss. Z. d. TH Karl-Marx-Stadt (1982), pp. 633 - 639.

[6] Mamdani, E. H. and T. J. Procyk: A linguistic self-organizing process controller. Automatica (GB) 15 (1979), pp. 15 - 30.

APPLICATION OF A FUZZY SET CONCEPT TO THE DESIGN OF CHEMICAL SYNTHESES

Weise A. and Kühnel W. [1)]

1. Introduction

Organic chemistry investigates the structures, properties and reactions of molecules containing the element carbon.
For communication and information about chemical structures chemists use the so-called structure formulas which represent the field specific language of chemistry. A structure formula is an iconic model of the true three-dimensional structure of a molecule.

Chemical reactions are expressed by structure formula equations, where the arrow shows the direction of the structural change (fig. 1). The arrow separates also the educt or input structures from the product or output structures of the reaction.

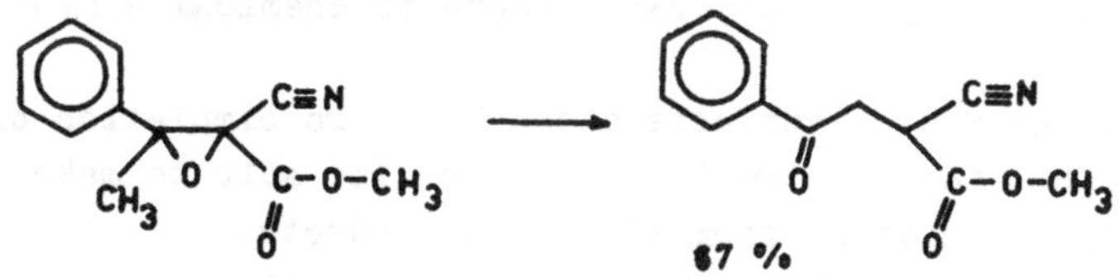

Fig. 1 Example of a chemical reaction equation

A very important additional information in the reaction equation is the yield in per cent. According to side and retro reactions nearly none reaction has a yield of 100%. A good yield is 80%, but for many reactions the yield is about 10 to 50%.
Therefore chemical reactions may be considered as <u>fuzzy relations by nature</u>.

An important characteristic of a reaction step is that only a substructure is incorporated in the structural change. These substructures are termed synthons, and the equation of changing substructures or synthons is the synthon transformation or simply transformation.
Fig. 2 shows the synthon transformation eliminated from the reaction equation in fig. 1.

Fig. 2 Example of a synthon transformation.
The starred atoms are the atoms prior connected to the other part of the complete structures.

---

1) Academy of Sciences of the GDR / Central Institute of Organic Chemistry, 1199 Berlin, Rudower Chaussee 5

The above mentioned conditions can be formalized by 2 relations:

A reaction is a relation between complete structures, dependent on some experimental conditions C and on a success value Y (1).

$$R = M^E \dashrightarrow M^P, C, Y \qquad (1)$$

A transformation is a relation between substructures, dependent on some limits with respect to free bonds F (at the starred atoms, s. fig. 2) and environment substructures ENV (2).

$$T = S^E \dashrightarrow S^P, F, ENV \qquad (2)$$

2. The CRDS Expert System

We developed an expert system for synthesis design which is based on relation (2) in his simulation part. The name CRDS (Chemical Reaction Design Software) points to the fact, that the system can be implemented in a flexible way for several demands of chemical synthesis design.

The system is able to generate synthesis plans by retro simulation of reactions backwards from target structures. It is also able to make predictions for forward reactions from educts to products.

Fig. 3 shows an example of a synthesis plan. The simulation goes downwards from the target to the precursors. The real syntheses are going up from the precursors to the target structure.

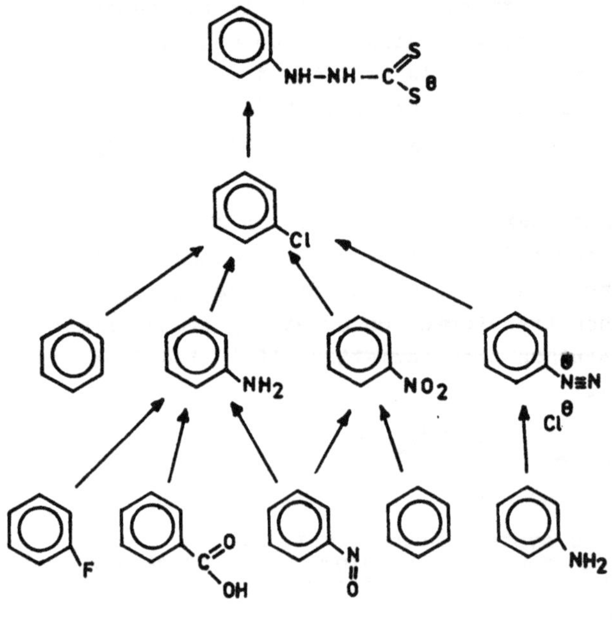

Fig. 3 A simple synthesis plan

The CRDS system is only partially constructed along the lines which are typical for the development of expert systems.
Normally the knowledge base of expert systems is introduced by experts who read the literature, extract the most important and most general facts and transfer them to the system.
The transfer process is often supported by a user-language tool, but nevertheless the process is tedious and time-consuming.

This obstacle is one of the reasons, why the automated extraction of knowledge from data bases is an essential goal of the Japanese 5th Generation Computer Project.
We realized such a feature in our CRDS system by a two-level learning function. Our system shows expertness without having had any contact to any expert.

The source of information in the CRDS system is a data base of reaction equations.
On the first learning level the synthon transformations are recognized. This is the task of a very complicated program. The transformations are carried over to the transformation library (TLIB) passing the second learning level (s. fig. 4).

In connection with the TLIB the <u>principle of synthon substitution</u> allows the simulation of any reaction in any of both possible directions (s. fig. 5).

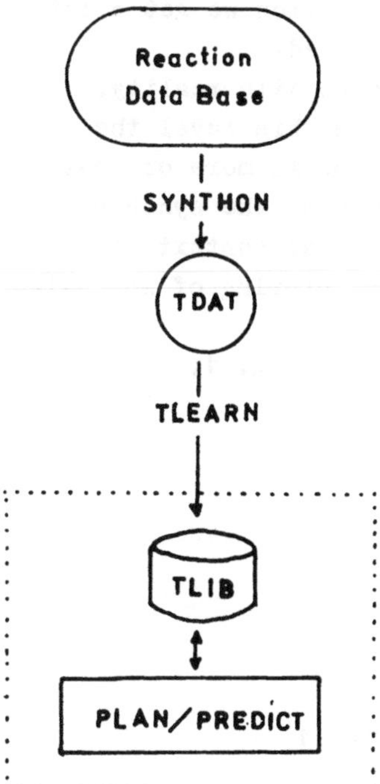

Fig. 4 Data flow in the CRDS expert system

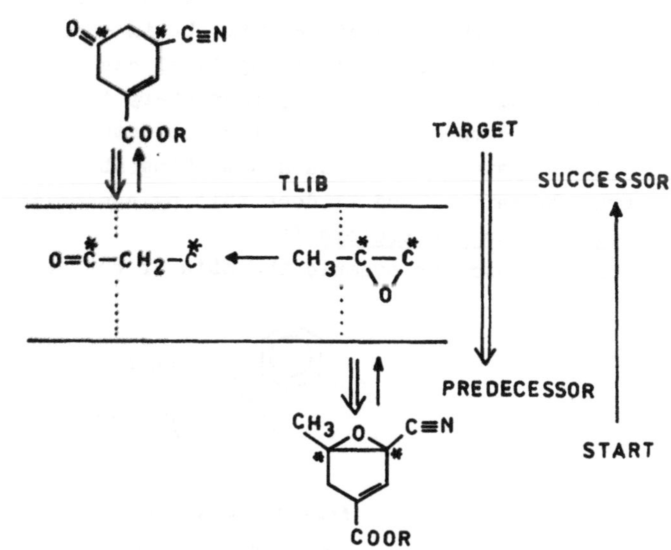

Fig. 5 Synthon substitution with the transformation of fig. 2

If we want to make synthesis planning the system has to look to the product side of the transformations in the TLIB, and it has to see, if any of the product synthons are included in the target structure.
In the positive cases the product synthons are eliminated and substituted by the educt synthons of the transformations. In this way precursors are generated.

For synthesis prediction vice versa the system has to look to the educt side of the transformations and to substitute them by the product synthons.

Obviously the performance of the system is dependent on the abundance of transformations in the TLIB.
The transformations are the production rules of the system, formulated in the language of chemistry and gained automaticly from a data base.

3. Fuzzy Synthon - Environment Relations

By formal using the principle of synthon substitution we get a lot of reaction proposals if the TLIB is reasonably filled.
Not all of these solutions will be in correspondence with reality.
Therefore the second learning level has to work. On this level the system takes into account, that every transformation is more or less coupled to the environmental substructures surrounding the synthons.
So the system has to learn something about the typical context of each transformation, and it does this by assembling and storing of so-called <u>environment descriptors</u> (ED).
Fig. 6 shows for example the ED's of the reaction in fig. 1.

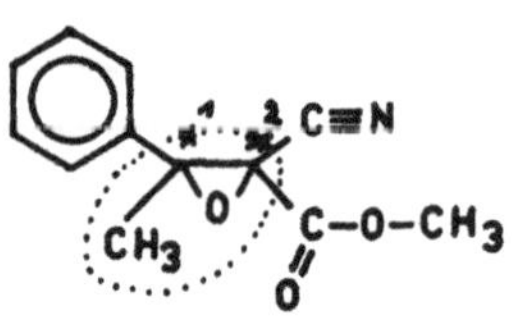

free bond $*^1$: Aryl
" " $*^2$: Acc, Acc

Fig. 6 The ED's of the example in fig. 1

The ED's primarily represent the types of specific substructure units. Because in the rule a transformation has been derived from several reactions, the frequencies of the ED's are also preserved. So at the first time the set of ED's per transformation was defined by relation (3):

$$ED_t = \{(g_{11}|f_{11}), \ldots\ldots (g_{nm}|f_{nm})\} \quad (3)$$

g = substructure group descriptor
f = frequency
n = number of groups per free bond
m = number of free bonds per transformation
t = transformation number

But experience shows that decisions based on ED's representing only type and frequency information are too sharp and lead to exclusion of realistic transformations.
This effect now shall be diminished by introducing fuzziness using the natural yield values associated with all reaction equations.
Every yield results from the combined effect of the environmental groups in a specific reaction. Therefore it is not possible to estimate the individual influence per environmental group and per transformation But it is possible to perceive a fuzzy yield range for every environmental group in a specific transformation.
The range is denoted by min, max-values and is derived from the yields accompanied with the different reactions belonging to the same transformation.
With this extension relation (3) proceeds to relation (4):

$$ED_t = \{(g_{11}|f_{11}|(ymin, ymax)_{11}), \ldots\ldots (g_{nm}|f_{nm}|(ymin, ymax)_{nm})\} \quad (4)$$

The arithmetic means of the $(ymin, ymax)_{ij}$ -pairs may be interpreted as fuzzy quantities marking the average influence of the environmental groups on the transformation:

$$\mu_t^m(g_{ij}) = \frac{(ymin + ymax)_{ij}}{200} \quad (5)$$

Moreover the employment of the ymin, ymax - values themselves by fuzzy union and fuzzy intersection for evaluation of the reaction steps in synthesis simulation (s. fig. 3) is very useful [1].
If synthon substitution takes place the actual set of ED's is a subset

of the ED_t set gathered in the TLIB for the transformation t.

$$ED_a \subset ED_t \tag{6}$$

Now we have the opportunity to evaluate an actual reaction step by three different membership functions (7) - (9):

$$\mu_a^p(t) = \mathrm{MIN}\ \left(\frac{ymin_b}{100} \right) \tag{7}$$

$$\mu_a^o(t) = \mathrm{MAX}\ \left(\frac{ymax_b}{100} \right) \tag{8}$$

$$\mu_a^m(t) = \frac{\dfrac{\sum \mu^m(g_b) \cdot f_b}{b}}{\dfrac{\sum \mu^m(g_c) \cdot f_c}{c}} \cdot \frac{1}{m} \tag{9}$$

b = number of groups in the ED_a set
c = " " " " " ED_t "
m = number of free bonds

In function (9) the average yield degrees and frequencies of groups in the actual reaction case are related to the whole ED set of the transformation as well as to the number of free bonds.
From function (7) results the most pessimistic, from function (8) the most optimistic membership degree of the actual reaction. There is an analogy to the degrees of nondominance and feasibility, defined by Orlovski [2].

The implementation of all three membership functions for the evaluation of reaction alternatives in the CRDS system is under way.

References

[1] Nishiwaki Y.: Possible Application of Fuzzy Set Concepts for Risk Analysis and Risk Assessment, Introductory Lecture, Joint IAEA/IIASA Risk Assessment Project, Division of Nuclear Safety, IAEA, Vienna, Austria (1983-03-16).

[2] Orlovski S.A.: Mathematical Programming Problems with Fuzzy Parameters, Working Paper 84-38, IIASA Laxenburg, Austria

A FUZZY CLASSIFICATION DIAGNOSIS SYSTEM

S.F. Bocklisch, F. Bilz [1], W. Schüler [2]

1. THE DIAGNOSTICAL PROBLEM

From a cybernetical point of view every real world system can be described by the automata equations. They give the relation between input signal $\underline{x}_e(t)$, output $\underline{x}_a(t)$ and state $\underline{x}(t)$, all time dependent and vectorial variables. Equations (1) and (2) describe the transfer characteristic and the system evolution by means of the operators G and F and their parameter vectors.

$$\underline{x}_a(t) = G\,[\,\underline{x}_e(t),\ \underline{x}(t),\ t,\ \underline{p}_G\,] \tag{1}$$

$$\underline{x}_{new}(t) = F\,[\,\underline{x}_{old}(t),\ \underline{x}_e(t),\ t,\ \underline{p}_F\,] \tag{2}$$

In that general form they describe dynamic, nonlinear and time-variant systems. In the diagnostic practice we have such ones. The diagnostical problem is to determine changes in G, F and $\underline{x}$. In that sense it is a type of instationarity. We have to remark that dynamic and timevariance is relatively. In most systems we observe both. They have only different time horizons.

The general diagnostical problem is to determine

a) the actual system situation (condition), we call that the <u>static diagnostic problem</u>

b) the time dependent change of a situation (looking for a trend), we call that the <u>dynamic diagnostic problem</u>, that is a problem of system evolution in a given environment.

The second consideration is neccesary for prediction and both exercises were done under risk.

The international trend leads to complex safety, diagnostic, inspection or consultation systems. They have some typical features like integration of sensors with fast signal processing, use of multivariate observations, use of all types of metric and nonmetric scales, that are proportional, interval and above all nominal and ordinal scales (linguistic variables - expressions of our normal and professional language), flexible communication devices like graphic input and output devices or language

---

1) Academy of Sciences of the G.D.R., Institute for Mechanics, DDR-1199 Berlin, Rudower Chaussee 5

2) District Hospital of Karl-Marx-Stadt, Radiologic Clinic and Policlinic, DDR-9003 Karl-Marx-Stadt, Bürgerstr. 2

analyser and synthesizer. Such systems have features of machine or artificial intelligence. They were called expert, computer assisted, computer aided or consultation systems and were implemented on powerful computers or near the process on process, personal or microcomputers for using at the place of employment or integrated into industrial process control units.

It should be possible to use such systems

- as on-line or off-line open-loop consultation systems for diagnosis
- as on-line closed-loop systems within automatic control

We developed such a system and it has the following components:

a) the process under investigation

b) the knowledge base with a large data base and a system for organisation including a problem adapted classifier using a pattern concept

c) the expert or teacher, necessary in the training phase of the system, respectively a self learning algorithm

d) the user itself aided by dialogue devices.

The knowledge base is grounded on our concept (/1/)

- using classification and additional
- fuzziness, that means uncertainty of informations, of the classifier model and the results on several levels.

2. STRATEGY

Fig. 1 shows the main components of our diagnosis system. We start with a set of experimental results - the data base. Each object, may be a technical or a medical system or a complex process situation, is described by a feature vector. Features are

- measuring results obtaind by using metric scales or
- expressions (values) defined on nonmetric nominal or ordinal scales; typical are linguistic variables with values like "small", "medium" or "big"

Data about the system that is under investigation can be devided into a number of different cathegories that have different grades of fuzziness.

a) History of the system

The history of a system is given by informations in the case of technological systems about

- the new state of the device (for instance a machine)
- the long time history available from special notices about working conditions, working ranges, unusual observations.

In the case of medical diagnosis the informations are available

- from patient himself about his diseases, such in his family, operations, allergies or his manner of living (rate of smoking, stress, diabetes and so on) or
- from patient papers and were interpreted and weighted by the physician in a subjective manner.

All that informations are fuzzy ones but nevertheless they are a very good basis for diagnosis and therapy decisions. In the case of technological systems that are instructions for repairing a device, for modyfying control parametes like speed, feed rate or tourque.

b) Actual informations

More or less objective data about the system we get by a physical examination or subjective interpretation of objective informations (X-ray pictures, tomograms). But engineers can make mistakes, overlook or misinterpret details. In such cases the qualification of the expert plays an enormous part but knowledge is never absolute and total. Other informations

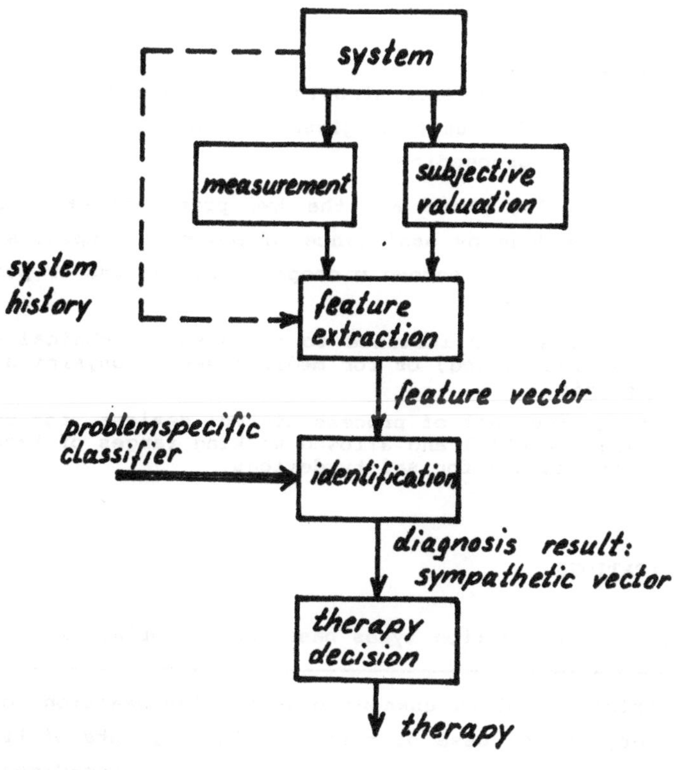

Fig. 1. Block scheme of the diagnosis system

are results of special tests, investigations for getting objective data on system characteristics, transfer functions, parameters, time constants and so on (see equations (1),(2)). However there exist measurement errors or artefacts and so we get imprecise and sometimes incorrect data caused by errors of measuring devices, blackouts, interrupts, ect.

Structure building

We use a classificational pattern concept. Clusters are defined

- by an expert, a teacher, as subjective given structures or
- as results of objective structure building procedures (cluster algorithms).

With every structure we define also a level of accurancy or uncertainty. For a special problem it is very important to define this level of view. So it is possible to get global or detailled reflections. Classifier structures of different roughness can coupled each together and we get a classifier net. So in a sequential way we can get diagnosis results with increasing accurancy.

Structure description

For description of the data structure we use the concept of a parametric membership function given in the paper of Bocklisch and Priber in this proceedings.

Learning and working phase are the two principal steps of the method. First is done by assistance of powerful computers in an off-line manner and for second microprocessorsystems or personal computer were applied

- near the process for instance to diagnose a technical system (like a machine device) or for medical use in physician practice or clinic
- as an integrated part of process control devices, for guaranty of optimal control and allowed working ranges of technological processes and for safety devices.

3. APPLICATIONS

Identification of reaction types based on vectorial time series

For identification of an unknown object often reactions of test functions or, in the case of periodic signals, cuts of time series are used. As an instance we take only an one-dimensional sampled time series that is the reaction curve of mean blood pressure of stressed rats. Each curve is described by 7 sampled values (every 5 hours, see fig. 2) and so we describe a test object by a 7-dimensional vector. Should we have other characteristical time series from the process, the vector would have 14, 21, ... components.

By hierarchical cluster analysis we got dependent on the threshold value, on different levels of roughness several cluster configurations. One is shown in fig. 2. There exist two typical reaction types. Now a special object is always a mixture of several types. So it is ingenious to use a fuzzy description in the time domain. Fig. 3 shows the description of every class by lines of constant membership values μ = const. = 1 (for the class representant), 0.75 and 0.5. For identification of an unknown object we have two possibilities (see the object in fig. 3 characterized by the dotted line):

a) a rough graphical identification by comparison the object curve with the fuzzy description of the reaction types

b) the discussion of the complete sympathetic vector.

It is seen that the given object has a higher affinity to type 1 and the sympathetic vector of $\underline{\mu}$ = (0.73,0.46) confirms that.

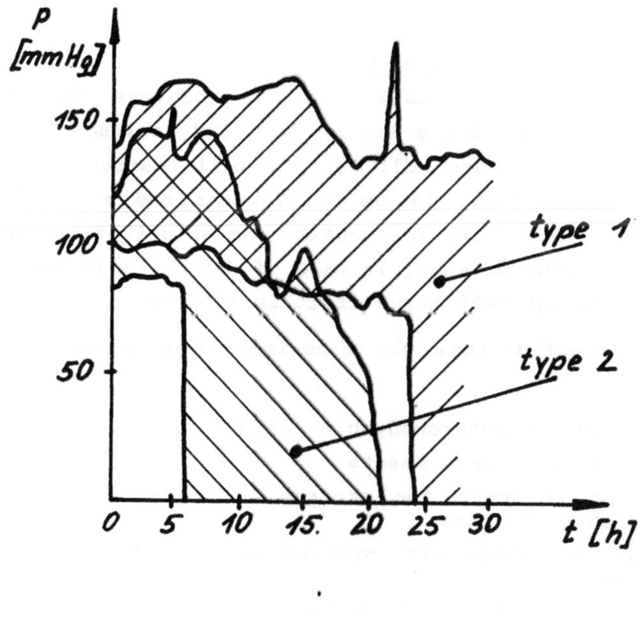

Fig. 2. Two reaction types of an ensemble of time series

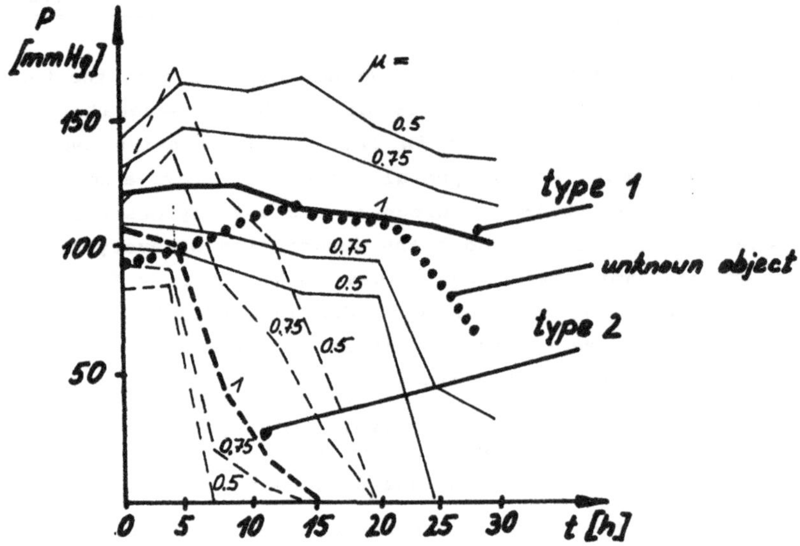

Fig. 3. Description of the two reaction types of fig. 2 by membership functions
(lines of membership μ = const. = 1, 0.75, 0.5)

Wear fault diagnosis of conveyor belt rolls

System under investigation was a roll of a conveyer belt used in a lignite mine. Such belts lie on socalled garlands consisting of 5 rolls each. Every roll has two ball bearings. It has dimensions of 60 cm long, 18 cm in diameter and a weight of about 30 kg. The problem is, to identify the actual state and to predetermine the trend of wear. In case of defect rolls they can become so hot that the belt burns if it is stopped.

For the test investigations the experimental arrangement consists of

- the test device for interchangable rolls
- sound and vibration measurements
- magnetic tape recorder for storage analoque signals with frequencies up to 20 kHz
- analoque filter devices with special determined characteristics (band passes)
- computation of effective values and
- a 8 bit microprocessorsystem for fuzzy classification including AD-converters.

Fig. 4 shows the result, a classifier with 4 classes given only in a two-dimensional feature space. The complete classifier is declared in a 8-dimensional space. Only the lines of constant membership value for μ = 0.5 are outlined.

There are two phenomena. First the normal wear effect, characterized by a movement of an object through classes 1 (new and normal ones), 3 (abraided ones) and 4 (bad ones) and second a defect of the bearing cover (class 2), a special device against dust and mad. The dashed lines are such for nonfuzzy decisions. They denote the valleys between two membership fuctions.

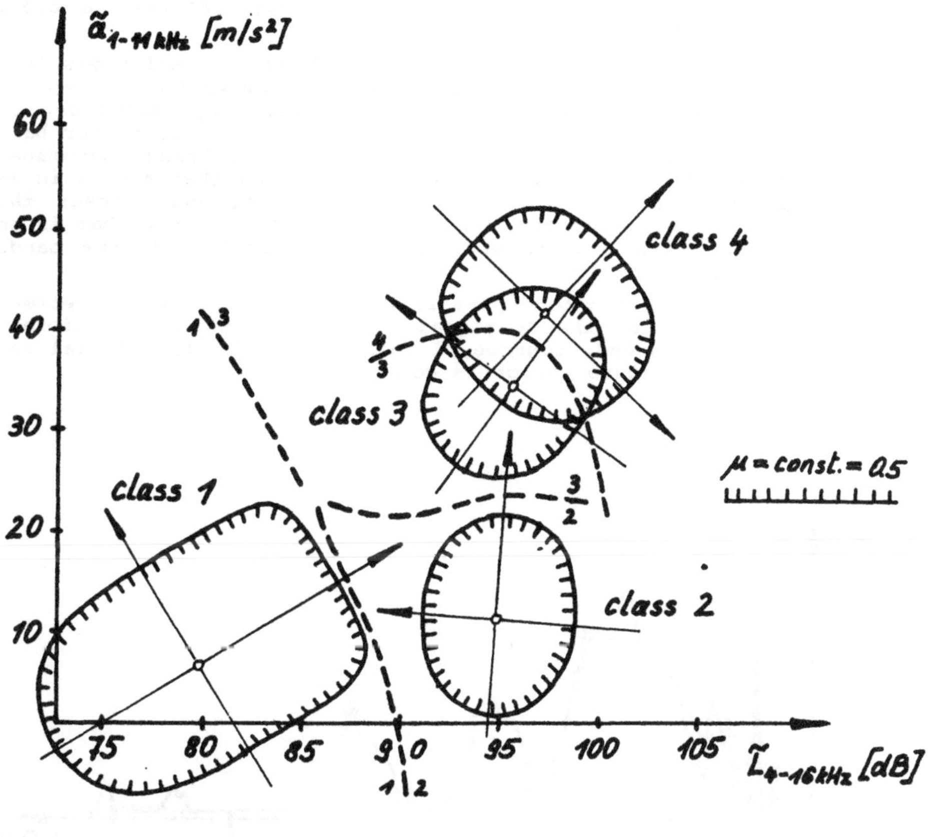

Fig. 4. Result of the four-class-configuration in the twodimensional feature space
(class membership functions are only characterized by lines μ = const. = 0.5)

Process inspection

Another problem we have under investigation is that of inspection of technical processes in machine tool industry. Especially the problem of identification of sharpness of drills and milling cutters in a numerical controlled working centre under working conditions. One of the difficulties is that the interesting vibro-acoustical signals, generated by the cutting process at the tool edges, and the measuring points at nonrotating machine parts have a relatively great distance each together.

The main steps for building and using a classifier are:

a) Choosing suitable measurement devices and measuring points and recording the analog signals with a magnetic tape recorder

b) Frequency analysis for getting a frequency spectrum for every test run (for instance boring a hole).

c) The averaged plots for a series of sharp and dull tools (see fig. 5) were interpreted by an expert. He discusses typical and untypical phenomena like peaks caused by vibrations of mechanical subsystems in resonance, speed and its higher harmonics, tourghe and also the interesting cutting phenomena. The expert determines the frequency regions that are of interest for this special inspection problem and chooses the characteristics of the filter devices. Sure, that that is an activity with uncertainty and that several different bands should be used.

d) Using the choosen band pass filters we get features as effective values of the AC-signal. Together with other informations like speed, tourque, length of the drill, material and so on the feature vector is completed.

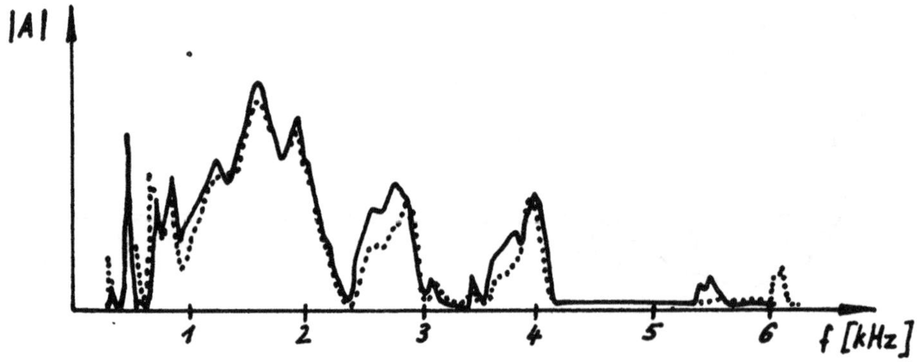

Fig. 5. Averaged spectra for sharp and dull drill tools

e) Now a classifier with hight classification quality is calculated with the aid of package EUKLID (/2/). Then it is implemented on a microprocessorsystem.

f) During the working phase the operator chooses the threshold for membership values. So the operator decision can be controlled by the actual requirements. It is evident, that it is impossible to work with very sharp tools everytime under conditions of high automatically production but it is necessary to hold the given quality standards.

Consultation system for mamma carcinom diagnosis

During last years for supporting the medical decision process manifold investigations have carried out to utilize computer assisted systems for making a diagnosis or differential diagnosis in almost all medical branches. The most frequent mathematical methods therefor are logical or probabilistic techniques. Also for supporting the medical decision special methods (predicate logic, Bayes-theorem, discrimination analysis) were applied to the interpretation of X-ray pictures, but did not find common application. Causes therefor are:

- almost only qualitative features
- missing existence of independence of the features
- big number of features
- subjective nature of the features due to visual interpretation of X-ray pictures by the physician
- for application of statistical theory marginal conditions are not performed
- big number of classes
- high uncertainty of the diagnostic decision at weak feature presentation

Most of the diseases require a high certainty of the diagnostic decision already in an early stage, because the therapeutic strategy and the prognosis of the patient depends on hereof in a high degree, especially for tumor patients. The medical check-up with X-rays (mammography) is the method of choice for early recognition of the mamma-carcinoma. The changes of the characteristics in the X-ray picture at early stages of mamma carcinoma, however, are very unspecifically. So a coordination to very much diagnostic classes is possible. In the international literature investigations at early stages show an identification probability of 65-70 percent for the physician. The application of logical and statistical methods does not improve this probability as investigations in our patients showed. We developed a special computer assisted decision system for early recognition of mamma carcinoma using the concept of the uniform parametric membership function on a personal computer system (see /3/). Characteristic parameters are

- 18 features
- 15 natural classes representing 6 semantic diagnostic classes
- possibility of the construction of classifier nets.

The 6 semantic classes were taken from physician expert knowledge and they contain 39 subclasses:

class 1: normal finding
class 2: all sorts of mastopathy
class 3: all sorts of benign tumors
class 4: all sorts of malign tumors
class 5: fibrosis
class 6: all other subclasses (mastitis, tuberculosis, ect.)

By means of hierarchical cluster analysis 15 natural classes were isolated out of the learning phase. The evaluation of the results of the cluster analysis allowed to assign the natural classes to semantic ones directly.

Our program permits the identification of unknown patients by computing the sympathetic vector to 15 classes. The decision for a particular diagnosis is made by the physician according to the maximal sympathetic value. But he takes into account also the next highest membership values.

For the construction of the fuzzy classifier during the learning phase the data from 1819 patients were taken and their final diagnosis was known. In a test phase data from 1492 unknown patients were taken and we got a hit probability of more than 95 percent for the correct identification of early stages of malign tumors, wich were confirmed histologically. For all other above mentioned forms of sickness of the mamma the hit probability is higher than 98 percent.

This diagnosis system on the base of fuzzy classification for early cancer dedection of mamma is introduced practically and implemented on a personal computer. Up to now about 6000 patients were classified by means of that system.

REFERENCES

/1/ Bocklisch,S.F.: Experimentelle Prozeßanalyse mit unscharfer Klassifikation, Diss.B., TH Karl-Marx-Stadt (1982)

/2/ Programmsystem zur unscharfen Klassifikation, Nutzerhandbuch, Akademie der Wissenschaften der DDR, Institut für Mechanik, Karl-Marx-Stadt (1984)

/3/ Schüler,W.: Experimentelle Expertsysteme für röntgenologische Diagnostikaufgaben unter Verwendung eines unscharfen Systemkonzepts, Diss.B, TH Karl-Marx-Stadt (1985)

/4/ Fortschrittsberichte der GDR Working Group for Fuzzy Sets and Systems, Akademie der Wissenschaften der DDR, FB Mathematik und Informatik, Band I bis VI (1977-1983)

Possible Application of Fuzzy Set Theory to Nuclear Safety Analysis, Risk Perception and Nuclear Plant Siting Evaluation

Yasushi Nishiwaki[1)]

Abstract

On case of a complex system, it is often difficult to evaluate the failure probabilities from past occurrences. Technical judgement and estimation must be used when it is necessary to assume possible failure of components which never failed before. The uncertainties involved in such human judgement and in human factors associated with man-machine systems may not necessarily be probabilistic, but fuzzy. Therefore, it would be important to develop a theory by which both non-probabilistic uncertainties, or fuzziness, of human factors and the probabilistic properties of machines can be treated consistently. From the mathematical point of view, probabilistic measure may be considered a special case of fuzzy measure. Fuzzy set theory seems to be an effective tool for analysing man-machine systems and large complex systems such as nuclear facilities. A concept "possibility of failure" based on fuzzy set theory or "fuzzy probability" which is a fuzzy set defined in probability space, is introduced. Possible application of fuzzy sets, fuzzy measures and fuzzy integrals to the analysis of nuclear risk perception and nuclear plant siting is also discussed.

I. INTRODUCTION

The Reactor Safety Study of WASH-1400 or the Rasmussen Report [1] was supposed to be realistic - as opposed to earlier studies that had been "worst-case" or "conservative" - and some uncertainties were also considered. The study was, however, subject to considerable praise and considerable criticism. Much of the criticism was directed towards the probabilistic and statistical methods used [2-4] . In fact, many of the data used in the fault-tree analysis of nuclear plants or nuclear fuel cycles are not the results of measurements but largely a matter of technical judgement and estimation. These estimates are then used in the presentation of the results as if they were frequencies in the probabilistic sense. The term "probabilistic" is sometimes applied to data in cases when it is not applicable, causing confusion in the interpretation of the result of some "probabilistic" risk analysis.

---

1)Division of Nuclear Safety, IAEA, Vienna, Austria

There are many types of uncertainties involved in safety analysis or risk assessment where biological or medical/societal or humanistic systems are concerned. However, they may be divided into two major categories: the uncertainty due to randomness and that due to fuzziness. Conventional methods of treating the uncertainty are to apply statistical methods of estimation which are, in turn, based upon the concept of probability. Even in the cases where the source of uncertainty is of a non-statistical nature, the formal application of statistical methods of analysis is often done to deal quantitatively with the uncertainty by tacitly accepting the premise that uncertainty - whatever its nature can be equated with randomness. Most of the works on safety analysis or risk assessment have been done using such methods. In the fuzzy set concept some uncertainties are accepted as uncertain with the introduction of the membership function. Instead of "true or false" of the non-fuzzy two-valued logic, any intermediate value between zero (false) and one (true) can be assumed for membership function in the fuzzy set theory proposed by Zadeh [5-15] .

A fuzzy set is a set of elements whose limit or boundary is not sharply defined but fuzzy. That is, the transition between membership and nonmembership of the elements to a given attribute is gradual and an element can "more or less" belong to the set, in contrast to the classical set, for which membership of the elements can be characterized by a binary form such as "yes" or "no", "true" or "false", "1" or "0", etc.

It has been pointed out by Nishiwaki [5,13,14] that the estimation of causal relationships between the initial damage and the end effects in radiation biology may be considered a problem of fuzzy inference, or fuzzy reasoning in fuzzy logic based on fuzzy target theory. The relations between the causes and the symptoms used in medical diagnosis are also treated as more or less fuzzy relations, because of the complex nature of human judgement and the medical, physiological and psychological factors involved [12] .

In this paper the problem of safety analysis will be discussed against this background. Such expressions as 'Low Probability/High Consequence', seem to be also a more or less fuzzy expression because 'how low is low', 'how high is high' and 'how safe is safe' are all fuzzy concepts and may to a large extent depend on human judgement or perception taking into consideration the various conditions and factors concerned.

The uncertainty due to vagueness of meaning or the uncertainty involved in human judgement or perception may not necessarily be probabilistic and could be treated with fuzzy set theory. In the fuzzy set theory the possibility measure is defined as distinct from the probability.

It was pointed out by Zadeh [6,12] : "what is possible may not be probable and what is improbable need not be impossible". A possibility distribution can be induced from a fuzzy set and does not underlie the idea of repetitive experiment, nor does a possibility measure satisfy the additivity property. If the possibility measures [6] are used in safety analysis, it may be called 'possibilistic' analysis [5,14,16] .

If a possibly very infrequent accident with high consequence is considered possible but practically improbable, it would be more appropriate to use the term 'low possibility' instead of 'low probability' [15] . According to the Oxford Dictionary, the first meaning of probability is 'the quality or fact of being probable'. If the frequency is extremely low, the term "improbability" or "impossibility" may be more appropriate.

II. HUMAN RELIABILITY AND THE FUZZY MODEL

In the past, for analysing the reliability of plants, probability theory has been applied in a way where an operator is considered as a machine component [19] . However, human performance is quite different from that of a machine in the following ways: it is (a) affected by many factors which are called 'performance shaping factors'; (b)adaptive, flexible and manageable for multi-jobs; (c) changed when partners work together.

Since many of these complex factors affecting human reliability performance may be considered more or less fuzzy, we need a theory by which both the vague property of human and probabilistic property of machine can be treated consistently. From a mathematical viewpoint, probabilistic measure is considered as a special case of fuzzy measure [18,20] . Therefore, fuzzy set theory can be an effective tool for analysing man-machine systems [21] . First we consider the linguistic expression of human reliability, which is more natural and more essential than a mathematical expression. This expression consists of a conditional statement, which is called a rule, and its truth value. That is

$$\text{'If A then B' is R} \tag{1}$$

where A is the causal event and B is the resultant event due to the human action. The truth value R, in this case, can be interpreted as the possibility of the transition from event A to B. That is to say, the conditional statement is a model of human behaviour, and the truth value R is a model of human reliability as shown in Fig. 1. A,B,R are usually all natural languages. In order to present Eq. (1) by fuzzy set theory, four types of model may be considered: (a) fuzzy relation, (b) fuzzy implication and truth value, (c) conditional fuzzy integral [19] and (d) fuzzy number and fuzzy function [6] .

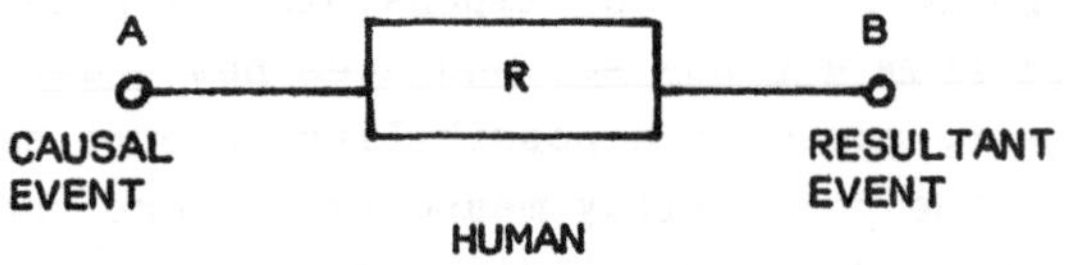

Fig. 1 Human reliability of single job

The last one is most simple and easily compared with probability. A fuzzy numbers is a fuzzy set of real numbers of which membership function is an isosceles triangle as shown in Fig. 2. The number corresponding to the peak is the most possible value and the length of the base means the vagueness.

Fuzzy function is a function of fuzzy numbers. By using fuzzy numbers and a fuzzy function, Eq. (1) is represented as follows:

$$B = F(A,R) \tag{2}$$

where F is a fuzzy function which is not different from an ordinary one. The calculation of fuzzy numbers is carried out by using the 'extension principle' as shown in Fig. 2 [6].

EXTENSION PRINCIPLE: $A*B = \sum \frac{\mu_A}{u_A} * \sum \frac{\mu_B}{u_B} = \sum \frac{\mu_A \wedge \mu_B}{u_A * u_B}$

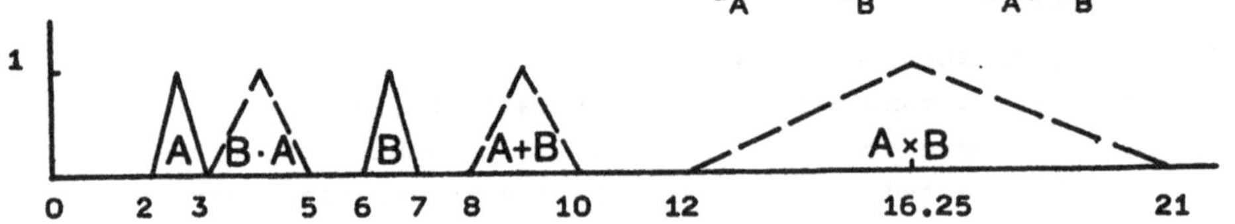

Fig. 2 Calculation of fuzzy numbers

Tsukamoto and Terano [22] and Terano et al. [23] have applied the notion of fuzzy logic to failure diagnosis of automobiles and diesel engines. The basic principle is to use all available information including the fuzzy information obtained by human operators to initiate a failure diagnosis [24,25]. In complex systems the early detection of any abnormal state, the diagnosis of cause and the suitable treatment are necessary for preventing accidents. The most important signs of trouble are possibly detected through the human operator's perceptions such as smell, noise, unusual indication of gauges or instruments, any unusual feeling or discomfort, etc. Diagnosis using this kind of fuzzy information seems to be more suitable than conventional information using deterministic non-fuzzy logic, especially when complex industrial processes such as nuclear facilities are in trouble [26]. Such fuzzy information as a slight burning smell could be an indication of immediate danger, but cannot be utilized effectively by an automatic safety control device based on non-fuzzy logic.

A new approach to fault-tree analysis and the diagnosis of abnormal events at nuclear power plants based on the fuzzy concept was presented by Sugeno, Onisawa, and Nishiwaki at the IAEA International Seminar, Dresden, 12-15 June, 1984 [16].

This paper introduces the concept of possibility instead of probability which is expresed using a fuzzy subset of unit interval [0, 1]. A fuzzy subset C, a fuzzy number, of [0,1] associated with a possibility distribution C(x) of the following type is considered.

$$C(x) = \frac{1}{1 + 20|x - c|^{m}} \qquad (3)$$

where,

$$c = f(\lambda_M)$$

$$\triangleq \begin{cases} \dfrac{1}{1 + \left(\frac{1}{3} \log \frac{1}{\lambda_M}\right)^3} & 0 < \lambda_M \leq 1 \\ 0 & \lambda_M = 0 \end{cases} \qquad (4)$$

This translation is deeply relaced to the logical connectives of possibilities. In the above expression, the parameter m is related to fuzziness and c, called the centre of C, gives the maximal grade of (C(x), i.e., C(c) = 1. We call C "failure possibility" and x "likelihood of failure"; we may say that failure possibility is a fuzzified likelihood of failure.

Table 1 shows the relation between a likelihood of failure c and a failure rate λ_M as well as the classification of c and their qualitative expressions.

Table 1 Classification of "c" (likelihood of failure)

Class	c	Qualitative expression	M
A	1.0	certain	1.0
B	0.9 - 1.0	most probable	3.0×10^{-2} - 1.0
C	0.7 - 1.0	quite probable	5.5×10^{-3} - 3.0×10^{-2}
D	0.5 - 0.7	probable	10^{-3} - 5.5×10^{-3}
E	0.3 - 0.5	racher improbable	10^{-4} - 10^{-3}
F	0.2 - 0.3	improbable	10^{-5} - 10^{-4}
G	0.1 - 0.2	highly improbable	10^{-7} - 10^{-5}
H	0.05- 0.1	extremely improbable	10^{-8} - 10^{-7}
I	0.0 - 0.05	excluded as får as one can judge next to impossible	0.0- 10^{-8}
J	0.0	impossible	0.0

Usually the failure rate of a component of a system is given by a triplet (λ_L, λ_M, λ_U) where λ_L is the lower bound of failure rate, λ_M the mean and λ_U the upper bound. In eq. (3), $m = m_L$ for $x < c$ and $m = m_R$ for $x \geq c$. The values of m_L and m_R are induced from k_L and k_R, where $k_L = \lambda_M / \lambda_L$ and $k_R = \lambda_U / \lambda_M$.

The values of m for different values of k and for different classes of "c" are given in Table 2.

Table 2 Values of m for k and for different classes of "c"

class	$k \leq 3$	$3 < k \leq 5$	$5 < k \leq 10$	$10 < k$
B	2.35	2.95	3.5	4.1
C	2.8	3.5	4.2	4.85
D	2.3	2.85	3.3	3.85
E	2.0	2.5	3.0	3.5
F	1.6	2.0	2.4	2.8
G	1.2	1.5	1.8	2.1
H	0.9	1.15	1.35	1.55
I	0.65	0.8	0.95	1.1

Fig. 3 shows typical examples of failure possibility according to its class.

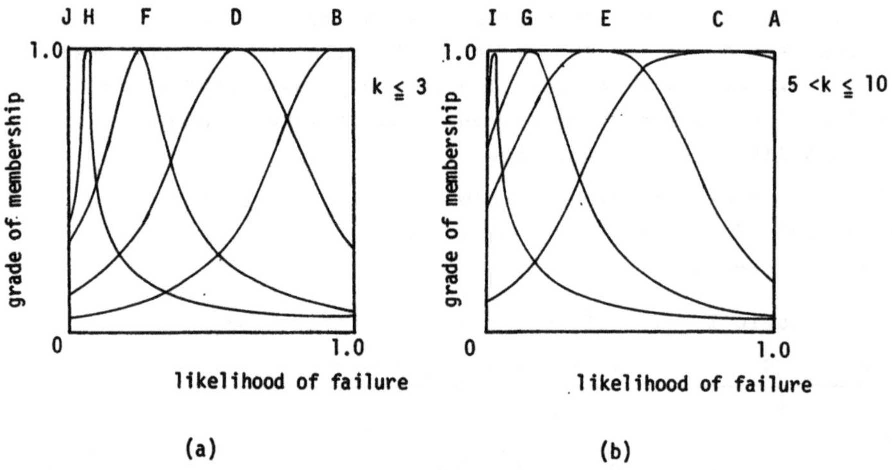

Fig.3 Examples of failure possibility

Sugeno, Onisawa and Nishiwaki used the concept "failure possibility" based on fuzzy sets to make safety analysis and fault diagnosis of a large complex system. Using failure possibility, a procedure to uniformly and consistently deal with component failure, human factors, environmental effects and common mode failure was shown.

A standby redundant system was also analysed by introducing fuzzy unreliability. As an example, the fault tree of a circuit diagram of power supply in a nuclear power plant was analysed, where the top event is a loss of electric power supply to engineering safety features [16].

III. FUZZY FAULT TREE ANALYSIS

In this section, the formulation of fuzzy fault tree analyses by Tanaka, et.al. is briefly reviewed [27]. A fault tree is a model that represents various logical combinations of possible events leading to the top event (undesirable event) of a system. An example is illustrated in Figure 4, where three types of event symbols and two types of gates are used. The rectangle defines an intermediate or top event as the output of a logic gate. The circle defines a primary failure of a system element, and the diamond represents a failure other than a primary failure, which is purposely not developed any further. The house does not represent a fault event but defines an ordinary event that inherently exists in the system. The circle, diamond and house are called fundamental events. The symbol "+" stands for an OR gate, and the symbol "." an AND gate.

In the example represented in Figure 4, the top event can be described as

$$T = X_3 \cup A_1$$
$$= X_3 \cup (X_1 \cap X_2) \qquad (5)$$

Given the failure probability of X_i, p_{X_i}, the failure probability of the top event T, p_T, can be written as

$$p_T(p_{X_1}, p_{X_2}, p_{X_3}) = 1 - (1-p_{X_3}) \cdot (1-p_{A_1})$$

$$= 1 - (1-p_{X_3}) \cdot (1-p_{X_1} \cdot p_{X_2}) \qquad (6)$$

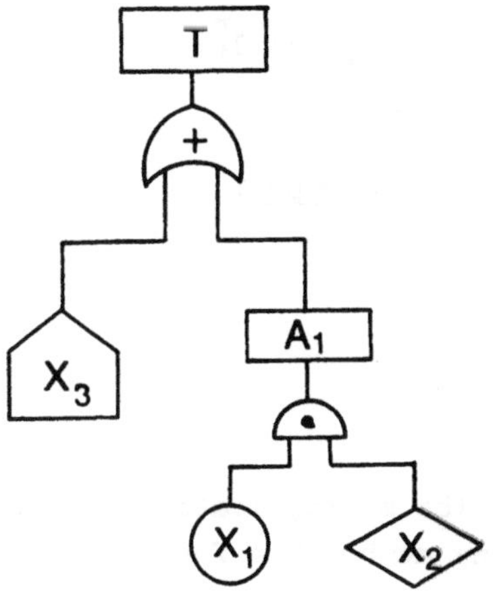

Fig.4 Illustration of a fault tree

The problem considered here is that of calculating the possibility of failure in the top event as a fuzzy set, given the possibilities of failure in fundamental events. Mathematically, it is equivalent to determining the following fuzzy set:

$$\underset{\sim}{P}_T(\underset{\sim}{P}_{X_1},\underset{\sim}{P}_{X_2},\underset{\sim}{P}_{X_3}) = 1 - (1-\underset{\sim}{P}_{X_3}) \cdot (1-\underset{\sim}{P}_{X_1} \cdot \underset{\sim}{P}_{X_2}) \tag{7}$$

where $\underset{\sim}{P}_{X_i}$ is a fuzzy set defined on $[0,1]$.

It is often difficult to assign a unique numerical value between 0 and 1 to a failure probability. To circumvent this difficulty, the failure probability may be defined as a fuzzy set on $[0,1]$. Specifically, the possibility of failure defined in a certain range on $[0,1]$ is used here instead of a unique value of probability. More specifically, the following type of fuzzy sets is considered in analyzing a fault tree.

The possibility of failure probability of X_i is denoted by

$$\underset{\sim}{P}_{X_i} = (q_i^{\ell}, p_i^{\ell}, p_i^r, q_i^r) \tag{8}$$

which is defined by the following membership function;

$$P_{X_i}(p) = \begin{cases} 0 & ;\ 0 \leq p \leq q_i^{\ell} \\ 1 - \left(\dfrac{p_i^{\ell} - p}{\Delta p_i^{\ell}}\right) & ;\ q_i^{\ell} \leq p \leq p_i^{\ell} \\ 1 & ;\ p_i^{\ell} \leq p \leq p_i^r \\ 1 - \left(\dfrac{p - p_i^r}{\Delta p_i^r}\right) & ;\ p_i^r \leq p \leq q_i^r \\ 0 & ;\ q_i^r \leq p \leq 1 \end{cases} . \tag{9}$$

where

$$p_i^{\ell} = p_i^{\ell} - q_i^{\ell} \tag{10}$$

and

$$p_i^r = q_i^r - p_i^r \tag{11}$$

According to the extension principle, the multiplication of fuzzy sets $\underset{\sim}{P}_{X_i} \cdot \underset{\sim}{P}_{X_j}$ gives rise to the following membership function:

$$\mu_{\underset{\sim}{P}_{X_iX_j}}(p) = \begin{cases} 0 & ; 0 \le p \le q_i^\ell q_j^\ell \\ 1 - \dfrac{h_{ij}^\ell}{2} + \sqrt{\dfrac{p - p_i p_j}{\Delta p_i^\ell \Delta p_j^\ell} + \left(\dfrac{h_{ij}}{2}\right)^2} & ; q_i^\ell q_j^\ell \le p \le p_i^\ell p_j^\ell \\ 1 & ; p_i^\ell p_j^\ell \le p \le p_i^r p_j^r \\ 1 + \dfrac{h_{ij}^r}{2} - \sqrt{\dfrac{p - p_i^r p_j^r}{\Delta p_i^r \Delta p_j^r} + \left(\dfrac{h_{ij}^r}{2}\right)^2} & ; p_i^r p_j^r \le p \le q_i^r q_j^r \\ 0 & ; q_i^r q_j^r \le p \le 1 \end{cases} \quad (12)$$

where

$$h_{ij}^\ell = \frac{\Delta p_i^\ell p_j^\ell + \Delta p_j^\ell p_i^\ell}{\Delta p_i^\ell \Delta p_j^\ell} \quad (13)$$

and

$$h_{ij}^r = \frac{\Delta p_i^r p_j^r + \Delta p_j^r p_i^r}{\Delta p_i^r \Delta p_j^r} \quad (14)$$

Since the procedure involved in the multiplication of two fuzzy sets is complicated, the following approximation may be used:

$$\underset{\sim}{P}_{X_i} \odot \underset{\sim}{P}_{X_j} \triangleq (q_i^\ell q_j^\ell,\ p_i^\ell p_j^\ell,\ p_i^r p_j^r,\ q_i^r q_j^r) \cong \underset{\sim}{P}_{X_i} \cdot \underset{\sim}{P}_{X_j} \quad (15)$$

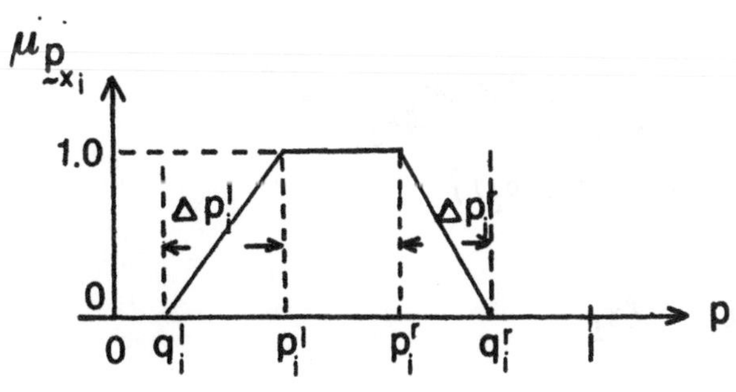

Fig.5 Fuzzy probability $\underset{\sim}{P}_{x_i}$.

Furthermore, we need the following exact calculations in evaluating the fuzzy set $\underset{\sim}{P}_T$, equation (7);

$$1 - \underset{\sim}{P}_{X_i} = (1-q_i^r,\ 1 - p_i^r,\ 1 - p_i^\ell,\ 1 - q_i^\ell) \quad (16)$$

Let us consider the example in Figure 4 with the following fuzzy probabilities;

$$\underset{\sim}{P}_{X_1} = (0.05,\ 0.1,\ 0.2,\ 0.24)$$

$$\underset{\sim}{P}_{X_2} = (0.06,\ 0.08,\ 0.1,\ 0.15)$$

$$\underset{\sim}{P}_{X_3} = (0.005,\ 0.01,\ 0.02,\ 0.025)$$

Since

$$\underset{\sim}{P}_T = 1 - (1 - \underset{\sim}{P}_{X_3}) \odot (1 - \underset{\sim}{P}_{X_1} \quad \underset{\sim}{P}_{X_2}),$$

we have

$$\underset{\sim}{P}_T\ (\underset{\sim}{P}_{X_1}, \underset{\sim}{P}_{X_2}, \underset{\sim}{P}_{X_3}) = (0.008,\ 0.018,\ 0.040,\ 0.060).$$

Some definitions are needed to compare fuzzy probabilities.

The fuzzy set

$$\underset{\sim}{P}_M = \max\ (\underset{\sim}{P}_{X_i}, \underset{\sim}{P}_{X_j}) \tag{17}$$

is defined by

$$\mu_{\underset{\sim}{P}_M}(p) = \max\{ \mu_{\underset{\sim}{P}_{X_i}}(p_i) \wedge \mu_{\underset{\sim}{P}_{X_j}}(p_j)\} \tag{18}$$

$$\{ p = \max\ (p_i, p_j)\}$$

Similarly, the fuzzy set

$$\underset{\sim}{P}_m = \min\ (\underset{\sim}{P}_{X_i}, \underset{\sim}{P}_{X_j}) \tag{19}$$

is defined by

$$\mu_{P_m}(p) = \max\{ \mu_{\underset{\sim}{P}_{X_i}}(p_i) \wedge \mu_{\underset{\sim}{P}_{X_j}}(p_j)\} \tag{20}$$

$$\{ p = \min\ (p_i,\ p_j)\}$$

If

$$\min\ (\underset{\sim}{P}_{X_i}, \underset{\sim}{P}_{X_j}) = \underset{\sim}{P}_{X_i} \tag{21}$$

or

$$\max\ (\underset{\sim}{P}_{X_i}, \underset{\sim}{P}_{X_j}) = \underset{\sim}{P}_{X_j} \tag{22}$$

the order relation is defined as

$$\underset{\sim}{P}_{X_i} \underset{\sim}{\leq} \underset{\sim}{P}_{X_j} \tag{23}$$

The above relation is illustrated with an example in Figure 6.

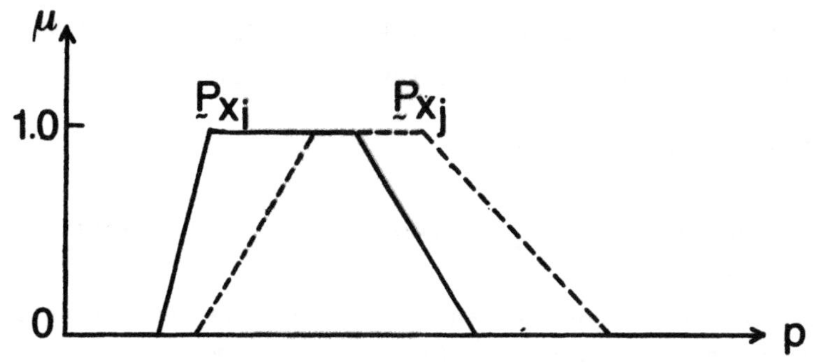

Fig.6 An example of $\underset{\sim}{P}_{x_i} \lesssim \underset{\sim}{P}_{x_j}$

By using the order relation, we may decide which fuzzy probability would be worse, that is if

$$\underset{\sim}{P}_{X_i} \lesssim \underset{\sim}{P}_{X_j} \tag{24}$$

then we should consider the event X_j to be more serious than the event X_i.

The following order relation

$$\underset{\sim}{P}_T(\underset{\sim}{P}_{X_1},\dots,\underset{\sim}{P}_{X_{i-1}}, 0, \underset{\sim}{P}_{X_{i+1}},\dots, \underset{\sim}{P}_{X_n})$$

$$\lesssim \underset{\sim}{P}_T(\underset{\sim}{P}_{X_1},\dots, \underset{\sim}{P}_{X_i},\dots, \underset{\sim}{P}_{X_n}) \tag{25}$$

may hold, since the probability of top event, $P_T(p)$, is an increasing function with respect to the probability p.

Suppose that

$$\underset{\sim}{P}_{X_i} \lesssim \underset{\sim}{P}_{X_j} \tag{26}$$

To measure the difference between $\underset{\sim}{P}_{X_i}$ and $\underset{\sim}{P}_{X_j}$, the following index may be introduced:

$$V(\underset{\sim}{P}_{x_i},\underset{\sim}{P}_{x_j})=(q_j^{\ell}-q_i^{\ell})+(p_j^{\ell}-p_i^{\ell})+(p_j^{r}-p_i^{r})+(q_j^{r}-q_i^{r}) \tag{27}$$

For simplicity, let us define

$$\underset{\sim}{P}_T(\underset{\sim}{P}_{X_1},\dots,\underset{\sim}{P}_{X_i},\dots,\underset{\sim}{P}_{X_n}) \triangleq \underset{\sim}{P}_T \tag{28}$$

$$\underset{\sim}{P}_T(\underset{\sim}{P}_{X_1},\dots,\underset{\sim}{P}_{X_{i-1}}, 0, \underset{\sim}{P}_{X_{i+1}},\dots,\underset{\sim}{P}_{X_n}) \triangleq \underset{\sim}{P}_{T_i} \tag{29}$$

Then, the value of $V(\underset{\sim}{P}_{T_i}, \underset{\sim}{P}_T)$ indicates the improvement attained by eliminating the possibility of failure of X_i. For example, if

$$V(\underset{\sim}{P}_{T_i}, \underset{\sim}{P}_T) \geq V(\underset{\sim}{P}_{T_j}, \underset{\sim}{P}_T) \quad (30)$$

preventing the failure of X_i is more efficient than removing the failure of X_j.

Let us consider a hypothetical example of the fault tree of a simplified numerical example in Figure 4. We have

$$V(\underset{\sim}{P}_{T_1}, \underset{\sim}{P}_T) = V(\underset{\sim}{P}_{T_2}, \underset{\sim}{P}_I) = 0.066$$

$$V(\underset{\sim}{P}_{T_3}, \underset{\sim}{P}_T) = 0.059.$$

Since

$$V(\underset{\sim}{P}_{T_1}, \underset{\sim}{P}_T) = V(\underset{\sim}{P}_{T_2}, \underset{\sim}{P}_T) \geq V(\underset{\sim}{P}_{T_3}, \underset{\sim}{P}_T),$$

the improvement due to the prevention of the failure of X_1 or X_2 is greater than that of X_3. Thus, if possible, the failure of X_1 or X_2 should be prevented instead of X_3in this case.

The following order relation may hold between the approximate product $\underset{\sim}{P}_{X_i} \odot \underset{\sim}{P}_{X_j}$ and the exact product $\underset{\sim}{P}_{X_i} \cdot \underset{\sim}{P}_{X_j}$.

$$\underset{\sim}{P}_{X_i} \cdot \underset{\sim}{P}_{X_j} \leq \underset{\sim}{P}_{X_i} \odot \underset{\sim}{P}_{X_j} \quad (31)$$

which means that

$$\max (\underset{\sim}{P}_{X_i} \cdot \underset{\sim}{P}_{X_j}, \underset{\sim}{P}_{X_i} \odot \underset{\sim}{P}_{X_j}) = \underset{\sim}{P}_{X_i} \odot \underset{\sim}{P}_{X_j} \quad (32)$$

This implies that the approximate operator some-what overestimates the possibility of failure, giving rise to a greater margin of safety [6].

IV. Fuzzy measures and Fuzzy integrals

(1) Fuzzy measures

According to Kruse [28], there are two fundamental mathematical concepts on fuzziness; one is a fuzzy set by Zadeh [7], as briefly explained in the introduction, and the other is a fuzzy measure by Sugeno [20] and Terano and Sugeno [18].

A fuzzy measure is an extended probability measure in one sense, which assumes, in general, monotonicity instead of additivity.

Let X be a universal set an $\mathcal{B}$ be a Borel field. Then a set function g defined on $\mathcal{B}$ with the following properties is called a fuzzy measure.

(i) $g(\phi) = 0$, $g(X) = 1$.

(ii) If $A, B \in \mathcal{B}$ and $A \subset B$, then $g(A) \leq g(B)$ (33)

(iii) If $F_n \in \mathcal{B}$ for $1 \leq n < \infty$ and a sequence $\{F_n\}$ is monotone (in the sense of iclusion), then $\lim_{n \to \infty} g(F_n) = g(\lim_{n \to \infty} F_n)$.

A triplet $(X, \mathcal{B}, g)$ is called a fuzzy measure space, and g is called a fuzzy measure of $(X, \mathcal{B})$. Note that a probability measure Pr has the properties:

(i) $0 \leq Pr(E) \leq 1$ for $E \in \mathcal{B}$, in particular $Pr(\phi) = 0$, $Pr(X) = 1$

(ii) If $E_i \in \mathcal{B}$ and $E_i \cap E_j = \phi$ for $i \neq j$, then $\sum_i Pr(E_i) = Pr(\bigcup_i E_i)$ (34)

From these it follows that a probatility measure satisfies the properties of a fuzzy measure.

For application, it is enough to consider a finite case. Let K be a finite set $K = \{s_1, s_2, \ldots, s_n\}$ and p(K) be a class of all the the subsets of K. Then a fuzzy measure g of (K, P(K)) is characterized by the first two propertis since the third one implies continuity. In particular, $g(\{s_i\})$ for a subset with a single element s_i is called a fuzzy density like a probability density. We denote $g^i = g(\{s_i\})$. Many types of fuzzy measures have been investigated since Sugeno suggested.

Fig. 7 shows the classification of fuzzy measures by Dubois and Prade [12].

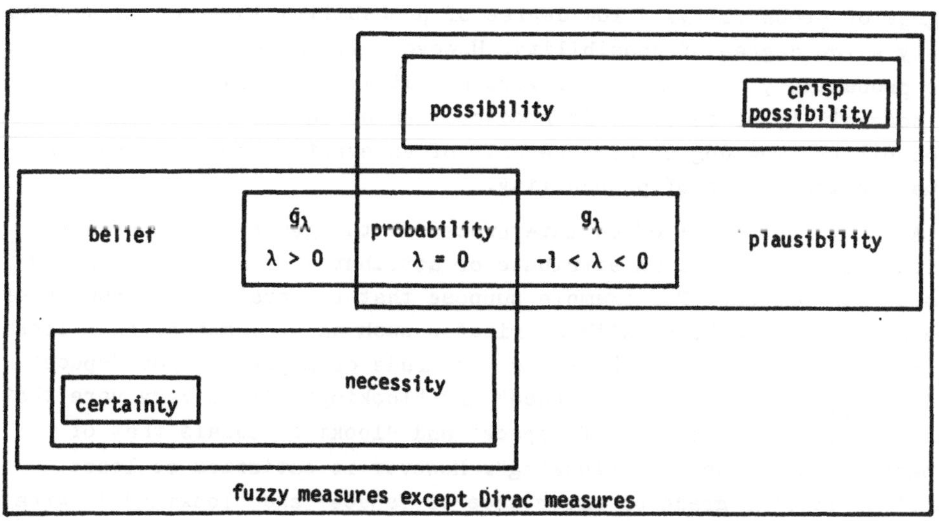

Fig.7 Classification of Fuzzy Measures [Dubois and Prade,12]

The probability measure P may be considered a special form of fuzzy measure. The relation (ii) in the above is called additivity and if $A \cap B = \emptyset$, then $P(A \cup B) = P(A) + P(B)$. The fuzzy measure approach by Sugeno generalizes probability measures by dropping the additivity property and replacing monotonicity. The possibility measure [6], the belief measure, the plausibility measure [23], etc. may be considered a kind of fuzzy measure. The fuzzy measure by Sugeno may be considered the most general concept of the measure in this respect. As a special form, g_λ measure has the following characteristics [12,20].

$$A \cap B = \emptyset \rightarrow g\ (A \cup B) = g_\lambda(A) + g_\lambda(B) + \lambda g_\lambda(A)\, g_\lambda(B) \qquad (-1 < \lambda < \infty) \tag{35}$$

When the fuzziness parameter $\lambda = 0$, it coincides with probability.

A possibility measure Π is a function which satisfies:

$$\begin{aligned} &\text{(i)}\ \ \Pi(\emptyset) = 0,\ \Pi(X) = 1 \\ &\text{(ii) For any collection } \{A_i\} \text{ of subsets of } X, \\ &\qquad \Pi(\cup_i A_i) = \sup_i \Pi(A_i) \end{aligned} \tag{36}$$

A possibility measure can be built from a possibility distribution, i.e. a function π from X to $[0,1]$ such that $\sup_{x \in X} \pi(X) = 1$ (normalization condition).
More specifically,

$$\forall A,\ \Pi(A) = \sup_{x \in A} \pi(X) \tag{37}$$

A high degree of possibility does not necessarily imply a high degree of probability. A low degree of probability does not necessarily imply a low degree of possibility. However, if an event is impossible, the probability is bound to be zero and it is improbable. The degree of possibility of an event is greater than or equal to its degree of probability. The degree of necessity of an event is smaller than or equal to its degree of probability.

A fuzzy measure is often interpreted as the grade of importance which one attaches to an attribute or attributes of an object in his evaluation process. For example suppose that the evaluation process of the niceness of a house with attributes such as s_1 : space, s_2 : looking, s_3 : price, etc.. Then $g(\{s_1\})$ is the grade of importance of "space" and $g(\{s_1, s_2\})$ is that of "space" and "looking". It is very rare that one's grade of importance of "space" and "looking" equals that of "space" added by that of "looking". However it satisfies at least monotonicity: the grade of importance of "space" and "looking" is greater than that of "space".

(2) Fuzzy integrals

Let h be a measurable function from X to $[0,1]$. Then the fuzzy integral of h over A with respect to g is defined as

$$\int_A h(x) \circ g = \sup_{a \in [0,1]} [\alpha \wedge g(A \cap F_a)], \tag{38}$$

where $F_a = \{x | h(x) \geq a\}$.

In the above, A is the domain of a fuzzy integral which is omitted if A is X.

Now let us see how to calculate a fuzzy integral. For simplicity, consider a fuzzy measure g of (K, P(K)) where K is a finite set previously defined.

Let $h : K \rightarrow [0, 1]$ and assume without loss of generality that $h(s_1) \geq h(s_2) \geq \ldots\ldots \geq h(s_n)$. Renumber the elements of K, if not. Then we have

$$\int h(s) \circ g = \bigvee_{i=1}^{n} [h(s_i) \wedge g(K_i)], \tag{39}$$

where $K_i \triangleq \{s_1, s_2, \ldots, s_i\}$.

A fuzzy integral can be used as a model of subjective evaluation of fuzzy objects [4] where the attributes of an object are measured by a fuzzy measure and the characteristic function of an object is integrated with respect to a fuzzy measure. In the previous example about a house, let $h : K \rightarrow [0, 1]$ be the characteristic function of a house, i.e., the function expressing the characteristics of a house.

For example, we set h ("space") = 0.9 for a big house and h("price") = 0.3 for a cheap house, etc.. Then the overall evaluation of a house is given by the fuzzy integral of h with respect to g, i.e., the grade of importance of each attribute.

In a fuzzy integral model, one does not have to assume the independency of an attribute from another. For example it is highly possible that there is a certain dependency between "space" and "price". If it is the case, we cannot use a linear model as far as we regard "space" and "price" as the attributes of a house. We have to consider the dependency of attributes from two points of view. One is objective dependency such as between "space" and "price", and the other is subjective dependency. Even if an attribute seems physically independent of another, one may consider that they are mutually dependent.

It should be also pointed out that a fuzzy integral is nonlinear in its nature. We do not have to assume that there hold additivity between the evaluation of one attribute and that of another attribute. As is well known, additivity is often a very ristrictive condition, for example, as is discussed in utility theory.

V. Conclusion and Discussion

The human subjective evaluation process has the following features.

(1) It is often pointed out that a man does not always have an additive measure such as probability to evaluate fuzzy objects.

(2) The attributes of an object in his evaluation process are not always independent of each other.

In either case, a linear model may not be applicable, since it is based on assumptions of additivity and independency.
Sugeno has proposed the concepts of fuzzy measures and fuzzy integrals and applied these concepts to model human subjectivity [20].
The concept "fuzzy measure" is considered as a generalized probability meausre, which is interpreted as a subjective measure to evaluate fuzzy objects. It is not necessary to assume subjective and/or objective independency and additivity in a fuzzy integral model. Using fuzzy integrals we can build a subjective evaluation model of a fuzzy object associated with various attributes.

In selecting the preferred candidates sites for nuclear power plants and for other nuclear facilities, it is essential to ensure that all site-related characteristics have been taken into account.

In evaluating the suitability of a site, various site characteristics have to be investigated and assessed; surface faulting,seismicity, subsurface material, volcanism, flooding, extreme meteorological phenomena, man-induced events, dispersion in air, dispersion in water, population distribution, feasibility of effective emergency actions, use of land in the region, availability of cooling water, etc. [30].

In addition to these site characteristics, public attitude or nuclear risk perception problems may have to be taken into consideration.

For the overall evaluation of various characteristics or attributes, the fuzzy integral model may be useful as demonstrated in the example of evaluation of a house. Onisawa, Sugeno and Nishiwaki, et al. applied fuzzy measures and fuzzy integral model to analyse nuclear risk perception or public attitude towards the use of nuclear energy based on the questionnaires prepared at the IAEA and the data collected in Japan [17].

The reliability of machine components is a measure of the capacity of the part of the machine to operate without failure when put into service. It is usually expressed with a probability associated with a non-failure performance of the device. In this case, reliability is the probability that a given system will perform as anticipated. The probabilistic approach is applied mostly to such cases as quality control or reliability of machine components, where the event or the test could be repeated many times under the same or similar conditions. However, systems reliability cannot be assessed by consideration of the hardware alone. The evaluation and prediction of potential human error is a most difficult problem, and therefore relatively little is known about predicting the performance and reliability of the man-machine interface.

This paper points out that the theory of fuzzy sets and fuzzy measures could be an effective tool for analysing man-machine systems for evaluating sefety in nuclear plant and siting.

In some of the nuclear risk assessment studies, fault trees and event trees are used to obtain models by which the probabilities of major accidents are expressed as functions of the probabilities of other events such as equipment failures and human errors. The uncertainties involved in technical judgement and estimation of input data are assumed as random variables sometimes without rigorous tests and treated with probability. However, if the chance of occurrence is extremely small and multiple repetition under the same condition is difficult, or if the human judgement or estimation is involved in the data without firm statistical evidence, the term possibility may be more appropriate to distinguish from probability [1-6, 14-16].

In some of the risk assessments of the population associated with nuclear accidents, the estimation is extended over a period of thousands or sometimes millions of years in the case of long-lived radionuclides in the environment.

However, in view of the fuzzy nature of various human factors and of unpredictable uncertainties of various conditions in the remote future, there must be some limitation (although it could be fuzzy) of time span, over which we could extrapolate our present-day knowledge with reasonable scientific accuracy. If we extrapolate it beyond this limit, we may be entering a field which is more fictitious than science fiction [15, 29].

In reality, sometimes, randomness and fuzziness are mixed. If the probabilistic approach for the analyses of uncertainties due to randomness and the new approach based on fuzzy sets and fuzzy measures for the analysis of uncertainties due to fuzziness or due to mixed causes are properly combined, it may provide us with an effective tool of systems analysis for safety of complex systems such as large scale complex nuclear facilities [15].

REFERENCES

[1] UNITED STATES NUCLEAR REGULATORY COMMISSION, Reactor Safety Study, WASH-1400, Office of Nuclear Reactor Regulation, Rep. NUREG-75/014 (1975) (Rasmussen Report).

[2] BJORK, L.: Problems of Risk Assessment, Statistics Sweden (Aug. 1983).

[3] SPEED, T.P.: "Negligible probabilities and nuclear reactor safety, another misuse of probability?", Department of Mathematics, University of Western Australia (1977).

[4] ESTERLING, R.G.: Statistical Problems in the Assessment of Nuclear Risks, Annual Meeting of American Statistics Association (Aug. 1980).

[5] NISHIWAKI, Y.: "Possible applications of fuzzy set concepts", Lecture of Risk Assessment Project IIASA/IAEA, 1983; "Risk assessment of atmospheric contamination due to combustion of fossil-fuels in Japan and possible application of fuzzy set", Proc. 10th Reg. Congr. IRPA, Avignon, France, Oct. 1982; "Possible application of fuzzy set concepts in radiation biophysics and safety assessment", Proc. 11th Reg. Congr. IRPA, Vienna, Austria, Sept. 1983.

[6] ZADEH, L.A.: Fuzzy sets, Inf. Control 8(1965)338-353; The concept of a linguistic variable and its application to approximate reasoning, Inf.Sci. 8(1975)199,301; 9(1975)43; Fuzzy sets on a basis for a theory of possibility, Int. J. Fuzzy Sets Systems 11(1978)3-28.

[7] KAUFMANN, A.: Introduction a la Theorie des Sous-ensembles Flous, Vol. 1-4, Masson, Paris (1972, 1975).

[8] NEGOITA, C.V., RALESCU, D.A.: Applications of Fuzzy Sets to Systems Analysis, Birkhäuser, Basel (1975).

[9] ASAI, K., NEGOITA, C.V.: Introduction to Fuzzy Systems Theory, Ohm Publ. Co., Tokyo(1978) (in Japanese).

[10] NISHIDA, T., TAKEDA, E.: Fuzzy Set and its Application, Kitamori Publ. Co., Tokyo (1978) (in Japanese).

[11] TERANO, T. (Ed.): Fuzzy Engineering, Kodan-sha, Tokyo (1982) (in Japanese).

[12] DUBOIS, D., PRADE, H.: Fuzzy Sets and Systems, Theory and Applications, Academic Press., New York (1980).

[13] NISHIWAKI, Y.: "Biophysical interpretation of the biological actions of radiations" (Proc. Int. Genetics Symp. Tokyo, No.271, 1956), Musashi Institute of Technology Press (1960); J. Radiat. Res., Chiba, Japan (1961)21, 22.

[14] NISHIWAKI, Y., et al.: "Possible applications of fuzzy set theory set theory in risk assessment, subjective perceptions and public attitutde study on nuclear energy", Proc. 6th Int. Congr. IRPA, Berlin, May 1984.

[15] NISHIWAKI, Y., et al.: "Accidents and Human Factors", Proc. Int. Symp. Jülich, 1984, IAEA, Vienna.

[16] SUGENO, M., ONISAWA, T., NISHIWAKI, Y.: "A new approach to fault tree analysis and diagnosis of abnormal events at nuclear power plants based on fuzzy concept", Proc. Int. Seminar, Dresden, 1984, IAEA, Vienna.

[17] ONISAWA, T., SUGENO, M., NISHIWAKI, Y., et al.: "Fuzzy Measure Analysis of Public Attitude Towards the Use of Nuclear Energy". Lecture of Risk Assessment Project IIASA/IAEA, 1984.

[18] TERANO, T., SUGENO, S.: "Conditional fuzzy measures and their applications", Fuzzy Sets and their Application (ZADEH, L.A., FU, K.S., TANAKA, K., SHIMURA, M., Eds), Academic Press, New York (1975) 151-170.

[19] SWAIN, A.D., GUTTMAN, H.E.: Handbook of Human Reliability Analysis with Emphasis on Nuclear Power Plant Applications, Battelle Columbus Labs, OH, Rep. NUREG/CR-1278.

[20] SUGENO, M.: Theory of Fuzzy Integrals and its Applications, Doctoral Thesis, Tokyo Institute of Technology (1974); Meas. Control 8(1972)218 (in Japanese).

[21] TERANO, T., MURAYAMA, Y., AKIYAMA, N.: Human reliability and safety evaluation of man-machine systems, Automatika 19(1983) 719-722.

[22] TSUKAMOTO, Y., TERANO, T.: "Failure diagnosis by using fuzzy logic", Proc. 16th Symp. Adaptive Processes, Vol. 2(1977) 1390-1395; Proc. IEEE Conf. Decision Control, New Orleans.

[23] SHAFER, G.: Mathematical Theory of Evidence, Princeton Univ. Press, Princeton, N.J. (1976).

[24] TSUKAMOTO, Y.: Fuzzy Logic based on Lukasiewicz Logic and its Application to Diagnosis and Control, Doctoral Dissertation, Tokyo Institute of Technology (1979).

[25] SANCHEZ, E.: Resolution of composite fuzzy relations equations, Inf. Control 30 (1976)38-48.

[26] SHAHIMPOOR, M., WELLS, D.J.: Applications possibilities for fuzzy failure analysis and diagnosis of reactor plant components and areas. Nucl. Eng. Des. 61(1980) 93-100.

[27] TANAKA, H., et al.:"Fault-tree analysis by fuzzy probability" IEEE Transactions on Reliability, Vol. R-32, No. 5, Dec. 1983.

[28] KRUSE, R.: "Fuzzy integrals and conditional fuzzy measure", Fuzzy Sets Systems 10, North-Holland, Amsterdam (1983)309-313.

[29] NISHIWAKI, Y.: Discussion at the IAEA Int. Conf. on Radioactive Waste Management, Seattle, 16-20 May, 1983.

[30] IAEA Safety Series No. 50-SG-S9: "Site Survey for Nuclear Power Plants", 1984, IAEA, Vienna.

Summary of the Panel Discussion on the Use of Fuzzy Set Theory to Nuclear Safety and Risk Assessment

Luis Lederman[1)]

The use of probabilistic methodology in evaluating nuclear power plants safety was briefly reviewed by Dr. Luis Lederman from the Division of Nuclear Safety of the International Atomic Energy Agency.

Steps leading to the determination of the risk curves based on a matrix formulation were introduced. According to this formulation, first used in the Zion Probabilistic Safety Study - USNRC Docket Nos. 50 - 295 and 50-304, a set of possible accident initiating events is initially selected. Next, accident scenarios are developed from each accident initiator. Binning of the various accident outcomes leads to the definition of plant damage states. The event three methodology is used to display the paths between the initiators and the plant damage states. Then a containment event tree is modelled to follow the phenomenology of the core meltdown and the possible failure modes of the reactor conteinment building to the eventual release of radioactivity to the environment. Thus, the entry states to the containment event tree are the plant states and the exit states are called release categories. Release categories are defined in terms of the quantity and mix of radioisotopes released, and other conditions like energy, high and duration of release. Finally, given that a particular release has occurred, a site model determines the consequent damage to people. For each particular release, the site model traces the movement of radioisotopes, the fall out and their interaction with the population. Meteorological and demographic conditions are considered in the modelling.

In mathematical terms the risk curves are depicted in terms of the product of a vector of accident initiators and three matrices, namely: a plant matrix representing the transitions from accident initiators to plant states, a containment matrix depicting the transitions from plant states to release categories; and a site matrix determining the consequence levels from release categories. Elements in the matrices represent the conditional frequencies of moving from a particular input condition to an output state.

[1)] Division of Nuclear Safety, IAEA, Vienna, Austria

The result of such an analysis is the determination of risk curves for each particular consequence index in a frequency format. These curves display the level of damage in terms of expected frequency of occurrence. The uncertainty in the curves are expressed by embedding them in a space of curves and constructing a probability distribution over this space. The degree of uncertainty depends on our total state of knowledge upon all the evidence, data, information and experience with similar courses of action in the past.

As of the question of uncertainities in probabilistic safety assessment (PSA) of nuclear power plants, three general categories were distinguished, namely:

(1) completeness uncertainties;
(2) modeling uncertainties and
(3) parameters uncertainties

The first refers to whether all significant phenomena have been considered. The second deals with the applicability of the models used to model component failures, human errors and physical phenomena. The third uncertainty refers to imprecisions in the input parameters utilized.

Uncertainties are treated in PSA studies by both Bayesian and classical methods. Bayesian processes have been used to include differences between sources of data, that is, variability of an event's rate of occurence from one facility to another. It was pointed out that uncertainties due to fuzziness are not distinguished in current PSAs.

The question was then raised on how to incorporate fuzzy logic in PSAs. It was clear, however, that the need exists to distinguish this class of uncertainties in particular when treating human errors and using engineering judgement or experts opinion.

It was concluded that need and ground exist to incorporate fuzzy logic in PSA studies. Although, specific steps to achieve these goals are not yet clear, it seems that a more general risk formulation would be needed to apply fuzzy logic.

It was suggested that a close cooperation between the Agency programmes in the field of probabilistic risk analysis and the other groups applying fuzzy set theory should be established for the benefit of Member States, and that a seminar could be organized to explore nuclear applications.

INFORMATIK · KYBERNETIK · RECHENTECHNIK

Herausgegeben von der Akademie der Wissenschaften der DDR

Zentralinstitut für Kybernetik und Informationsprozesse

Die Schriftenreihe INFORMATIK · KYBERNETIK · RECHENTECHNIK umfaßt alle Fachgebiete, die in der Hauptforschungsrichtung Kybernetik und Rechentechnik der Akademie der Wissenschaften der DDR integriert und Gegenstand der Vorlesungen und Forschungsaufträge an Universitäten und Technischen Hochschulen sind.

Folgende Themenkomplexe werden behandelt:

Automaten- und Algorithmentheorie

Informationstheorie und Automatisierung

Programmiersprachen und Rechentechnik

Datenerkennung und Datenverarbeitung

1

Dieter Bochmann/Christian Posthoff

Binäre dynamische Systeme

1981. VIII, 397 Seiten — 168 Abbildungen — 25 Tabellen — 17 × 24 cm
Lederin 68,— M (z. Zt. vergriffen)
Bestell-Nr. 7623619 (6391)

Aus dem Inhalt: Dynamische Vorgänge in binären Systemen — vorrangige Interpretation Schaltnetzwerke und Automaten — mathematische Beschreibung durch Booleschen Differentialkalkül — technische Anwendung bei rechnergestützter Analyse von Schaltnetzwerken

2

Siegfried Unger/Fritz Wysotzki

Lernfähige Klassifizierungssysteme

1981. XI, 294 Seiten — 77 Abbildungen — 8 Tabellen — 17 × 24 cm
Lederin 48,— M; Ausland 60,— M
Bestell-Nr. 7623563 (6372)

Aus dem Inhalt: Künstliche Intelligenz — psychologische Begriffslernexperimente — Klassifizierung von Objekt- bzw. Datenmengen — Wahrnehmungsprozesse — statistisch-orientierte Verfahren der Diskrimination von Verteilungen

3

Horst Völz

Information I

Studie zur Vielfalt und Einheit der Information. Theorie und Anwendung vor allem in der Technik

1982. 407 Seiten — 179 Abbildungen — 83 Tabellen — 17 × 24 cm
Lederin 78,— M; Ausland 98,— M (z. Z. vergriffen)
Bestell-Nr. 7620602 (6244/I)

Aus dem Inhalt: System und Information — formale Sprachen — Information in der Fernsprechtechnik — Speichertechnik — Optik — Sprache — Massenmedien — Psychologie — Physiologie — Informationstheorie — Versuch zur Klassifikation und Definition der Information

4

Horst Völz

Information II

Ergänzungsband zur Vielfalt und Einheit der Information. Theorie und Anwendung vor allem in der Biologie, Medizin und Semiotik

1983. XII, 370 Seiten — 184 Abbildungen — 60 Tabellen — 17 × 24 cm
Lederin 72,— M; Ausland 98,— M
Bestell-Nr. 7629690 (6244/II)

Aus dem Inhalt: Bedeutung des Informationsbegriffs in Biologie und Medizin — Genetik der Zellen — Nervensystem — Sehen — Hören — Gedächtnis — Kommunikation und Semiotik — Definition der Information

5

Programmsysteme

Anwendung — Entwicklung — Fundierung

Herausgegeben von Peter Bachmann

1983. X, 228 Seiten — 53 Abbildungen — 18 Tabellen — 17 × 24 cm
Lederin 52,— M
Bestell-Nr. 763 006 9 (6644)

Aus dem Inhalt: Programmsysteme als Hilfsmittel moderner Rechenanlagen — Architektur — Steuerung — mathematische Modellierung — Entwicklungs- und Nutzungstechnologie

6

Günter Riedewald/Jan Małuszyński/Piotr Dembiński

Formale Beschreibung von Programmiersprachen

Eine Einführung in die Semantik

1983. X, 205 Seiten — 37 Abbildungen — 16 Tabellen — 17 × 24 cm
Lederin 42,— M; Ausland 60,— M
Bestell-Nr. 763 136 0 (6720)

Aus dem Inhalt: Moderne Methoden der Programmierung — formalisierte Beschreibung von Programmiersprachen — Syntaxdefinition — Semantikdefinition — attributierte Grammatiken — algebraische Modelle von Programmiersprachen

7

Norbert Ahlbehrendt/Volker Kempe

Analyse stochastischer Systeme

Nichtlineare dynamische Systeme

1984. 335 Seiten — 37 Abbildungen — 7 Tabellen — 17 × 24 cm
Lederin 58,— M
Bestell-Nr. 763 122 1 (6714)

Aus dem Inhalt: Lineare Systeme und Markowsche Diffusionsprozesse — stochastische Differentialgleichungen — Analyse durch Funktionalreihenentwicklung — statistische Linearisierung — Volterra-Reihe aus dem Blockschaltbild — Momentenentwicklung der Verteilungsdichte — zeitliche Mittelungsmethode

8

Henning Schoch

Programmieren in PL 1

1984. 520 Seiten — 7 Abbildungen — 17 × 24 cm — Lederin 68,— M
Bestell-Nr. 763 135 2 (6719)

Aus dem Inhalt: Systematische Darstellung des vollen Sprachumfangs von PL 1 — zahlreiche Beispiele zur Veranschaulichung der Anwendungsmöglichkeiten einzelner Sprachelemente — Darlegung der wichtigsten Beziehungen zum Betriebssystem OS/ES

10

Helmut Löffler

Rechnerverbundsysteme

1984. VIII, 296 Seiten — 176 Abbildungen — 12 Tabellen — 17 × 24 cm
Lederin 54,— M; Ausland 68,— M (z. Z. vergriffen)
Bestell-Nr. 763 162 7 (6737)

Aus dem Inhalt: Architektur von Rechnerverbundsystemen — Mehrrechnersysteme — Rechnerkommunikationsnetze — Rechnernetzwerke — rechnergestützte Telekommunikationssysteme

Bei Fragen zur Produktsicherheit wenden Sie sich bitte an:
If you have any questions regarding product safety, please contact:

Walter de Gruyter GmbH
Genthiner Straße 13
10785 Berlin
productsafety@degruyterbrill.com